Survismeter

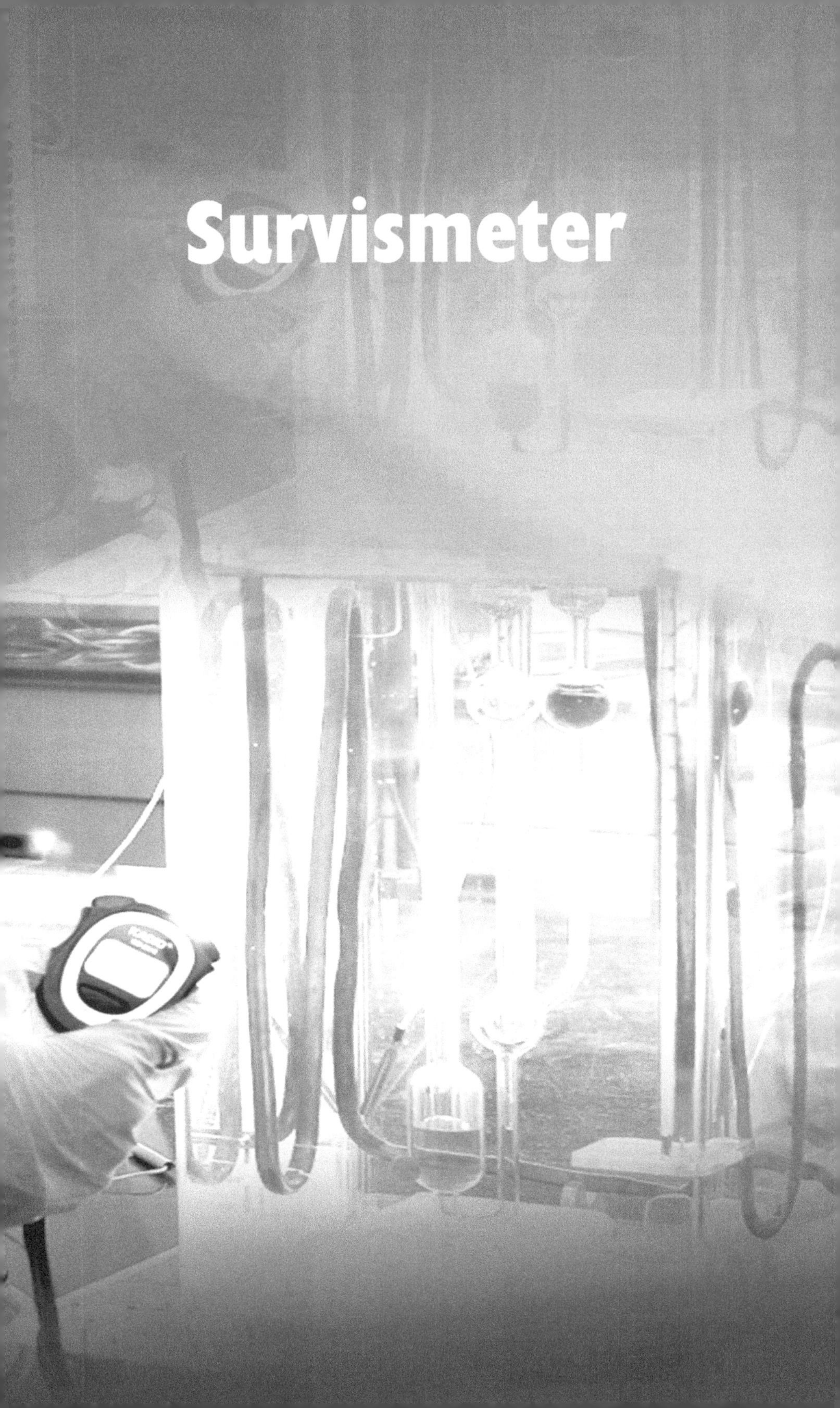

Survismeter

Fundamentals, Concepts, and Applications

Man Singh
Sunita Singh

PAN STANFORD PUBLISHING

Published by

Pan Stanford Publishing Pte. Ltd.
Penthouse Level, Suntec Tower 3
8 Temasek Boulevard
Singapore 038988

Email: editorial@panstanford.com
Web: www.panstanford.com

British Library Cataloguing-in-Publication Data
A catalogue record for this book is available from the British Library.

Survismeter: Fundamentals, Concepts, and Applications

ISBN 978-981-4774-70-3 (Hardcover)
ISBN 978-0-429-02761-1 (eBook)

Contents

Preface

The potential of a single molecule has become an unavoidable resource for use in several academic as well as industrial and analysis applications. Tracking molecular resources needs extraordinary research activities as well as technology. The molecule constitutes the functional, nonfunctional, intermediate, or conformational state or metastable materials. The molecule constitutes the object-making material on mixing several other molecules, like dyes or colorants, dispersants, solubilizers, wetting agents, and reflecting light. Materials like dyes, cosmetics, solvents, kitchenware, writing ware, glassware, computer ware, food and nutritional materials, paper and pulp, and others are created from binding and intrabinding activities of solvents, molecules for higher stabilities of industrial products like thin films, coating, dyeing, drugs, catalysts, adsorbent, with their most optimized molecular force vectors with higher potential energy on Lennard-Jones potential scale. Intermolecular binding on friccohesity and tentropy scales for such products is needed to optimize the residual forces contrary to the binding activities of the drug; and glucose, fructose, sucrose, maltose, proteins, starch (as a food molecule), hormones, and enzymes need nonbinding states for better functionality, performance, and efficiency. Thereby friccohesity, which moderates the binding as well as nonbinding activities quantitatively, defines their degree and physicochemical states. Such natural needs have generated or created a vacancy in nanoscience and nanotechnology where the molecules independently work with the targeted objects rather than among themselves. Such activities among themselves lead to the development of cohesive forces, and activities with other molecules induce intermolecular forces (IMFs). Such interdependence of cohesive forces as well as IMFs is called friccohesity. Also molecules like drugs, adsorbents, catalysts, and fluorescent dyes need to be in independent states for larger surface areas to perform tasks efficiently. Considering such domains of molecular interaction engineering the physicochemical properties (PCPs) of the molecular materials become useful and what is needed is their accurate measurements, interpretations, illustrations, modeling, and interconnections for product formulations. Such sciences are an indispensable part of material characterization, and hence tracking the physicochemical potential of molecules with the survismeter could be most useful for better product formulation out of experimental data. The survismeter seems to be a turning point in the area of physicochemical sciences. For fitting of fairly defined theories vis-à-vis the repercussions of the PCP data, the explanation of experimental data for developing new science and working model out of such simulations of the materials is needed urgently. And in the absence of adequate scientific resource materials this book gains significance in offering wider solutions and formulating new models.

The book is aimed to act as a lively foundational resource as well as a reference material in the area of nanomaterials. This book is a unique model of resource materials that deals with experimentation with the survismeter to generate accurate data on PCPs of formulations designed with several molecules, salts, ions, nanoparticles, and capped nanoparticles. Also several molecules need to be in an independent state in their nanoemulsion formulations, and this book excellently covers their fabulous science and tentropy.

Man Singh
Sunita Singh
Autumn 2018

Chapter 1

Introduction

Considering the needs and applications of physicochemical properties (PCPs) to elucidate the interacting mechanisms of molecules, molecular ions, and ions, the contents and subject matter of this book could prove to be a substantial resource and a masterstroke in areas of molecular signatures in day-to-day physicochemical processes in biochemical, biotechnological, biological, nanoscience and nanotechnology, food and nutrition, and biophysical areas under infinitesimal as well as noninfinitesimal variations in natural as well man-made processes. The book explains the role and significance of capillary science and its application in tracking or quantitative determination of PCPs of molecules and ions separately and together. It deals with electronic profiles and their effects on interactions and also ionic sizes and their effect on hydration spheres or ionic solvation, entropy, and viscosity as distribution processes of molecules. It very plainly explains the wettability and interfacial tension (IFT) of immiscible liquids and their mutual mixing with cationic, anionic, and nonionic surfactants along with dendrimers. It comprises 63 chapters, and each chapter provides a foundational explanation of the science in ascending and descending order or the structure making and breaking activities, respectively. It explains the mechanisms of micelle and reverse micelle formation,

Survismeter: Fundamentals, Concepts, and Applications
Man Singh and Sunita Singh

ISBN 978-981-4774-70-3 (Hardcover), 978-0-429-02761-1 (eBook)
www.panstanford.com

trapping of air, percolation of cations and anions in a central hydrophilic cone in the case of reverse micelles, and also the solubilization of a drug in the case of micelles where hydrophobicity versus hydrophobicity adjoins and establishes the contact or overlap of electron clouds for stability of the combinatorial interface or linkages for transportation and suitability as per need. Entropy of hydrophobic and hydrophobic combinatorial ratios decreases on increase of Gibbs energy. The molecular language scripted out of electronic configuration is an industrial process, and their signatures are defined with mathematical relations and coordination.

1.1 Survismeter: An Experimental Resource

The quadrupolar science of the virtual valve–driven survismeter has become an indispensable experimental resource for exploring solution and nanoemulsion interfaces of academic and industrial significance. For new hypotheses, system and process designing, implementation, reproducible data production, and interpretation, wet laboratory research initiates and develops new models as new state of art. In chemical- or molecular-science-driven research, the molecular makeup of materials is a most fascinating backbone of electronically configured atomic bonds where valence electrons in atoms within the Bohr radius of 5.3×10^{-2} nm play a deciding role in molecule making. However, looking to reveal the said molecular science with suitable devices becomes a most attractive reverse science. In chemical processes, especially liquid mixtures, enthalpy, entropy, tentropy, and potential energy vis-à-vis cohesive forces (CFs) and distribution forces of individual molecules are significant in the context of the intramolecular multiple force theory (IMMFT). So the solubility and dispersibility in chemical processes retrieve a chemical-makeup-driven interacting behavior in desired solvents or mixed solvents. The solvent becomes homogenized as equilibrium-distributed CFs are noted as indistinguishable global forces. All chemicals are not soluble or dispersible in all solvents. Solvents dissolve or disperse solutes in their volumes against the resistance of the solvent's hydrogen bonding (HB). Resistance causes a friction in the flow or dispersion of the solute or dispersate, which develops frictional forces (FFs) or distinguishable forces. So friccohesity rationalizing FFs and CFs offer the solutions in connivance with Debye–Hückel and Boltzmann distribution vis-à-vis van der Waals, Lennard-Jones potential (LJP), and London dispersive forces (LDFs). The study of solvent activities dissolving glucose, salts, and others in aqueous and trimesoyl tridimethyl malonate ester (TTDMM) dendrimer dispersion in a nanoemulsion is challenging. Surfactants disperse the oil in water, forming nanoemulsions. For such processes, physicochemical study needs to determine viscosity, surface tension, IFT, wetting coefficient, friccohesity, surface area, particle size, and activation energy. The survismeter—a dimension- and geometry-driven device—measures them. It works on the basis of degenerate canonical equilibrium and partition function with rigid boundary condition. The survismeter holding liquid helps determine drug binding and release, oil distribution, and solution interaction and build new databases. It retrieves and reveals electrostatic, dimerization, and intramolecular hydrogen bonds. It does not work on the basis of the relaxation model: time and space. Surface tension and viscosity explain in an integrated manner particle and activation energy distribution and partitioning in nanoemulsions. So in a nanoemulsion, rigid rotor, rotational, translational via hydrogen bond, electrostatic poles, and dipolar changes are estimated, bringing academic and industrial potential of liquids for advancing and creating new chemistry. In this book, studies conducted on the behavior of drugs, dyes, dendrimers, proteins, and DNA will be discussed, along with survismeter science and its applications.

1.2 Survismeter: A Flagship Science

The survismeter liquid mixtures experimental science and technology (LIMEST) is the biggest scientific and technological interface to track molecular interfacial potential to cater to one of the largest needs of basic and applied research. Molecular modeling needs configurational, combinatorial, and experimental data of molecules for better science and application of molecular as well as non-molecular materials. The LJP and density functional theory (DFT) or friccohesity functional theory (FFT) need advanced and deep insight into electronic distribution and wave functions generated therein via the molecular orbital approach through the highest occupied molecular orbital (HOMO) and the lowest unoccupied molecular orbital (LUMO) states of the molecules. Sigma bonds vis-à-vis the HOMO and pi bonds or electrostatic poles vis-à-vis the LUMO tract the LDFs and van der Waals forces, respectively, for interaction in the process. Potential interfaces, like the electrostatic molecular orbital state (EMOS), with definite friccohesity establish unique molecular expressions. The EMOSs that are generated from the HOMO and LUMO models lead to an unequal distribution of the changes that generate interacting abilities among the molecules, either with themselves or among the different molecules. Such renewed charge distributions may be in response to temperature, pressure, compositions, geometry, or the mechanical energy and induce vacancies for the emergence of friccohesity. As the interacting abilities between similar molecules cause CFs, the abilities between dissimilar molecules develop intermolecular forces (IMFs). CFs and IMFs both are the main domains of friccohesity. Therefore, such a unique breakthrough for predicting the interacting and reacting faces and interfaces of molecules becomes useful in sciences of thin film, coating, and wetting of the solid surface with a definite isotropy. Such approach of research and understanding of the overall mechanism of the molecules and ion molecules through electron spins, wave function, and entropy lays down the foundation for friccology. And friccology could furnish a broader electronic profile of internal arrays of vectors of molecular forces in the case of molecular interaction or capping, coagulations, and coalescence. Thus in a nutshell, friccology is the study of mutual sharing of more than two molecules in liquid mixtures. Friccology could assist in furnishing better explanations of shear stress and shear strains as functions of molecular structures, collectively called rheology. Thus friccology and rheology both are interconnected sciences where the friccologist studies the interactions of more than two molecules in liquid mixtures.

The interacting molecules develop several ordered and unordered zones with higher and lower entropies. The survismeter LIMEST shapes the future of the most dynamic materials or

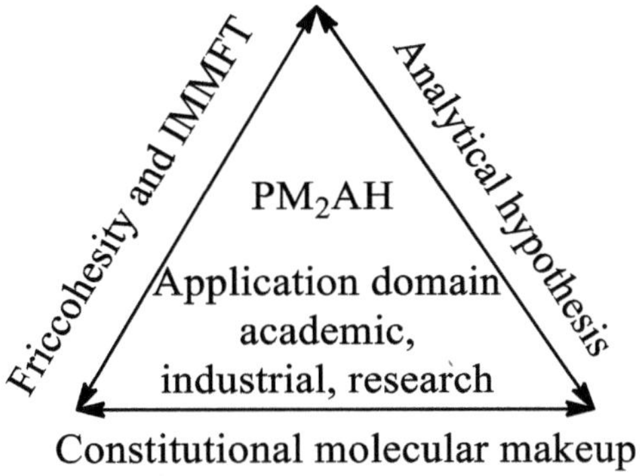

Figure 1.1 Survismeter produces authentic data.

molecules by measuring their physicochemical profiles for identifying their use in several disciplines. The survismeter LIMEST is emerging as one of the most informative flagships of world-class sciences, applications, and education, supporting world-class publications and applications of materials in the sciences wherever the formulations have become essential. The survismeter has emerged through a research paper submitted on June 16, 2003, and accepted on August 19, 2004 in *Instruments and Experimental Techniques* [1]. Currently thousands of users have been using the survismeter for their research works. The survismeter has been upgraded and is making a constructive impact on the careers of millions of scientists, students, entrepreneurs, and citizens worldwide. The survismeter, being highly sensitive, responds to electronic changes, present nature and positions of functional groups, double bonds, inductive effects, molecular geometries, activation energies, adsorption, chain length, and charge accumulations, in several solvents. The molecules, via various atomic oscillations along the bonds, are involved in molecular motions, which become cutting edge for fluid dynamics combated with survismeter science. There seems a constitutional linkage or a connection of PCPs with molecular makeup and analytical grilling or analytical hypothesis in constitutionalizing the PCP–molecular makeup–analytical hypothesis (PM$_2$AH) as a new dimension of survismeter science (Fig. 1.1).

1.2.1 Detection of Alcoholic (–OH) Functional Group

The survismeter detects the number of functional groups like –OH in the molecules. Figures 1.2 and 1.3 depict structural reorientations and surface tension measurements for aqueous D-sorbitol and aqueous glycerol at 293.15 K and 298.15 K, respectively. D-sorbitol contains six and glycerol three –OH groups. Glycerol, with three –OH groups, decreases the surface tension, probably because it weakens the CFs of the water and strongly interacts with the water monomers. The alkyl chain of glycerol of three carbon atoms also remains operational to cause hydrophobic interactions. In such structural engineering neither the hydrophilicity of the three –OH groups nor the hydrophobicity of the alkyl chain dominates.

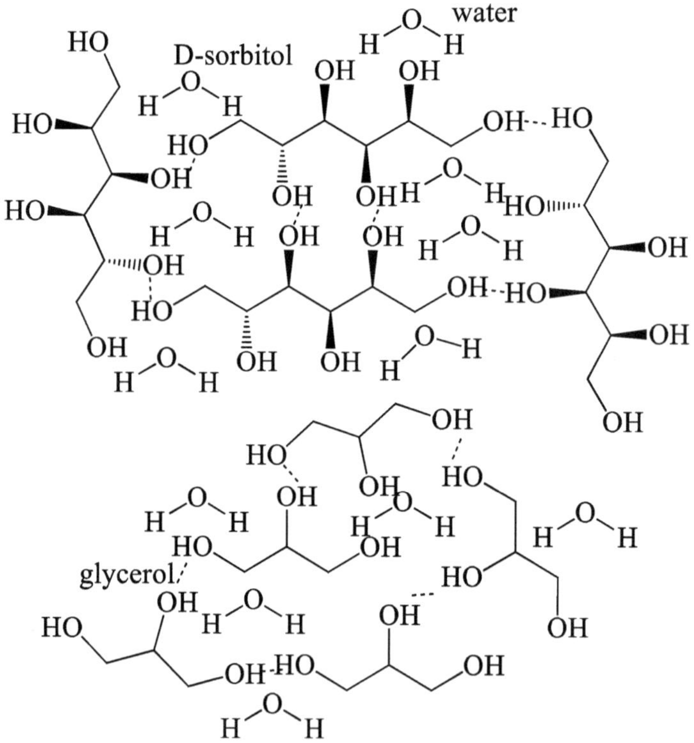

Figure 1.2 Structural reorientations for aqueous D-sorbitol and aqueous glycerol at 293.15 K and 298.15 K, respectively.

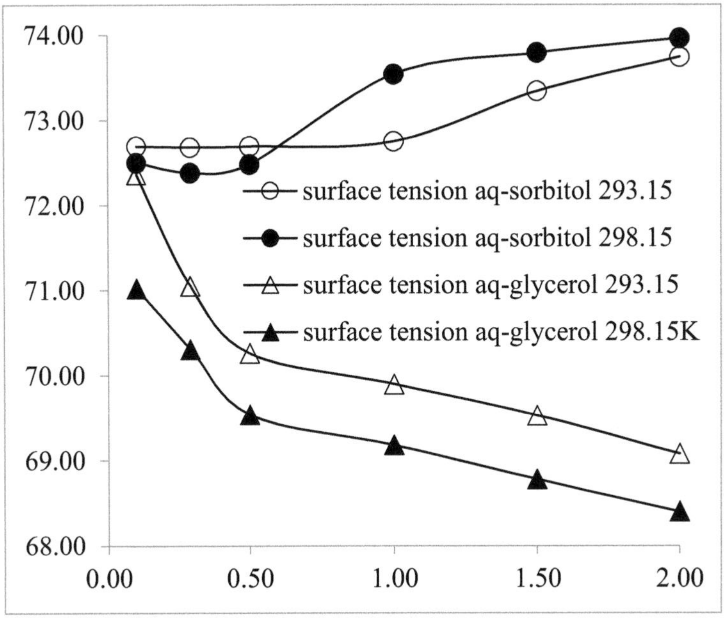

Figure 1.3 Glycerol (with three –OH groups) acts as a surfactant, but D-sorbitol (with six –OH groups) acts as a salt.

Contrary to such interaction scenario of glycerol with water the sorbitol, with six –OH groups, increases the surface tension, which shows that the stronger hydrophilic interactions of the six –OH groups dominate and develop stronger hydration spheres around them (Fig. 1.3).

Such hydration spheres tend to embed the –OH at their center, and ultimately the CFs again dominate among the water molecules. Such interaction behavior of the –OH numbers is most fascinating as the survismeter is capable of determining the science of the increasing numbers of alcoholic –OH groups. Till date no method could distinguish the effect of the number of alcoholic –OH groups. However, the study is continued with other functional groups. It is also a matter of research that a temperature increase of 5°C has crucially favored their respective effects.

Figure 1.3 depicts that glycerol (with three –OH groups) acts as a surfactant but D-sorbitol (with six –OH groups) acts as a salt.

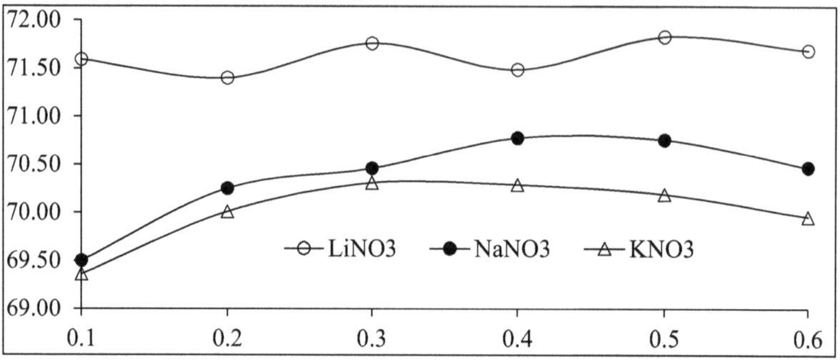

Figure 1.4 Surface tension of aqueous lithium, sodium, and potassium nitrate salts against their molalities.

It shows that a stronger hydrophilic acts as a salt but a stronger hydrophobic acts as a surfactant.

1.2.2 Detection of Cations

The higher the cationic size, the lower is the surface tension. Similarly, the larger the size of the cation, the higher is the ionic field around the ion, which strongly disrupts the structure of the water and applies stronger water-holding capacity with lower friccohesity. Such arrangements of smaller ions increase the surface tension and viscosity both, but large cations, though they attempt to disrupt the water structure, cannot develop a strong ionic field around themselves (Fig. 1.4).

Therefore, water is not strongly aligned and weaker CFs are generated, which induce lower surface tension and higher viscosities in the absence of weaker CFs. This clearly indicates the development of a stronger ionic field around the small cation with higher surface tension. It indicates that the ionic strengths vary as $Li^+ > Na^+ > K^+$, with the strengths of attraction forces between the oppositely charged hydrations of the cation and NO^- anion as $K^+ > Na^+ > Li^+$ for $2s^1 > 3s^1 > 4s^1$ orbitals, respectively. So Li^+ aligns dipoles of water with stronger CFs and hence $LiNO_3$ acts as surface tension and viscosity enhancer. Large-sized KNO_3 slightly decreases the surface tension and increases the viscosity with higher hydration–hydration interactions. Therefore, cations can be identified by measuring their surface tension and viscosity with the help of a survismeter. However, the effect of smaller cations on larger cations is not known and is also not reported, so their studies could be planned with the survismeter.

1.3 Survismeter: A Boon for Nanomaterials

Adequate and accurate physicochemical characterization brings to our focus the variable or the degree of freedom of chemical processes. The degree of freedom decides the stoichiometry combination with respect to the experimental conditions for developing greener routes of the industrial process for developing nanoemulsions. Currently, demands of the material-synthesis-characterization application (MCA) triangle as an integrated interface essentially needed for scientific modeling have occupied a centerstage for understanding, advancing, and developing new state of the art in scientific streams for better material development in areas of textile, dyeing and printing, cosmetics, sensors, graphene, food and drug, pesticides, syrups, and thermodynamically stable nanoemulsions. For such approaches and modeling a deep understanding of behavior of materials has become a unique requirement for the practical task of verification of the desired or thematic coordinates on practical scales. Such scaling needs experimental measurements and fitting the experimental data on a statistical task for accurate and correct Gaussian distribution analysis for scientific validity or approval. Further one or more than one atom join another atom or atoms where such joining occurs between similar atoms or different atoms with different electronic configurations and electron distribution densities, ultimately a bonding develops between the atoms, leading to a chemical bond with definite shared and unshared electron pairs that are covered within the LUMO and HOMO orbital sciences.

Molecular models with definite electronic densities develop electrostatic centers or dipoles in the molecular structures, which when they are brought in contact with the solvent induce interactions. Such electron-driven structural modification induces molecular activities vis-à-vis the molecular nature of the solvent. Of course, sigma bonds with the HOMO model also lead to structured

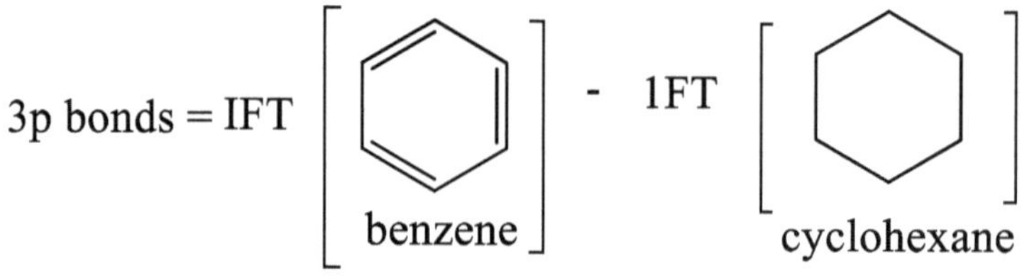

IFT of 3 pi bond = 14.32-10.59 = 3.59 mN/m

Figure 1.5 IFT values of benzene and cyclohexane determined with the survismeter.

hydrophobic interactions (SHI). The wonders of SHI vis-à-vis the LUMO are expressed in liquid phases, like the organic phase (CCl_4) with the water phase noted as immiscible liquid phases with definite IFT (mN/m). Figure 1.5 depicts the IFT values of benzene and cyclohexane determined with the survismeter. These values enable researchers to determine the contribution of pi bonds of molecules. No device other than the survismeter is available that determines the contribution of pi conjugations of organic compounds or solvents in terms of the IFT. It predicts the role of the three pi bonds of benzene for inducing mutual solubilizations of water and benzene together, which produces 10.59 mN/m IFT against 14.32 mN/m of cyclohexane. Thereby the LUMO and HOMO model generates a vacancy of measurements of IFT. The hydrophilic and hydrophobic activities of the surfactants affect the IFT values. Such science becomes useful when the secondary bonds are either disrupted or developed because of the state of electrostatic charges on the molecules with definite structure breaking and making (SBM) activities.

Similarly, the HB sites increase mutual solubilization of isopropyl alcohol and water-immiscible solvent systems on addition of ammonium acetate and sodium acetate salts. Sodium acetate, with a stronger ionic field with Na^+ and a weaker ionic field with the CH_3COO^- anion, develops higher IFT with weak mutual solubilization of water and isopropyl alcohol. However, ammonium acetate enhances their mutual solubilization because it has more HB sites for developing with the water structure. Thus, Na^+ strongly disrupts the water structure and aligns the water molecules for

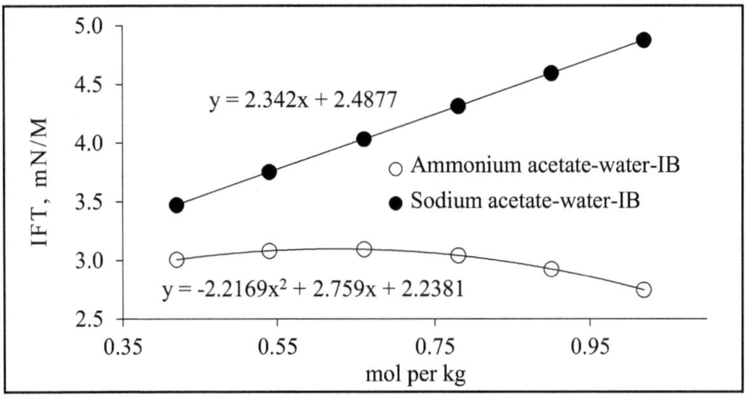

Figure 1.6 Interfacial tension of ammonium and sodium acetates-water with isobutyl alcohol.

developing stronger CFs with higher surface properties, producing higher IFT values. At the same time ammonium acetate favors their mutual solubilization with lower IFT values (Fig. 1.6).

SBM activities have been a fundamental step for designing, capping, and dispersion of NPs with a desired set of solvents with definite friccohesity. Such experiments need immediate attention for quantitative measurements of cohesion or CFs (surface energy or surface tension, mN/m) vis-à-vis dispersing or cohesion disrupting activities (viscous activities or viscosity, mPa·s) generating IMFs among the contents, activation energy, NP surface area, and size. These data help ascertain the state of the resultant liquid sample as an emulsion, a microemulsion, or a nanoemulsion. The magnitudes of the CFs are responsible for the emulsion, microemulsion, or nanoemulsion state, where the higher are the CFs the weaker are the oil-and-water-binding forces (OWBFs), which are responsible for nanoemulsion formation. Thus weaker OWBFs have a weaker mutual distributing ability and liquid mixtures in this state have lower viscosity or shear. Therefore, computational simulations of variations of cohesion and transformation into IMFs are initially tracked with the survismeter and are defined by friccohesity. Friccohesity is a fundamental breakthrough of recording and reviewing CFs as well as IMFs, which are solely responsible for nanomaterial formation or synthesis. Such simulatory sciences need

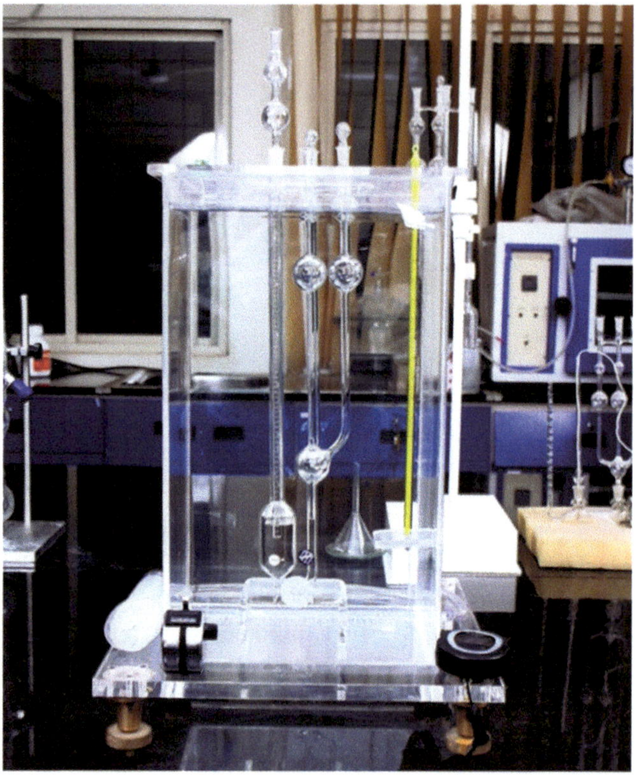

Figure 1.7 Experimental setup of the survismeter.

quantitative data of CFs and IMFs. The experimental setup of the survismeter is illustrated in Figs. 1.7 and 1.8.

1.3.1 Merits of the Survismeter

- It is a fast and multipurpose technology.
- Human error is eradicated in measurements.
- There is zero loss of sample in multioperational steps and so less sample requirement.
- It is ecofriendly and user friendly and produces authentic data.
- Fabrication expenditure is reduced.
- Glass-blowing gases (oxygen gas and LPG) and manpower requirement are reduced.

Figure 1.8 Experimental setup of the survismeter.

- Heat emission to the environment during glass blowing is reduced.
- It is safe to transport and maintain.
- There is a huge reduction in chemicals and solvents in conducting experiments.
- It promotes green chemistry and innovativeness among youngsters.
- Less time is involved in analysis.
- It is basic-maintenance-free and reduces the use of cleaning reagents.
- It reduces operational steps in measurement.
- The infrastructure requirement is minimized as compared to that of individual devices.
- It efficiently estimates capping and aggregation of nanoparticles (NPs) of gold, silver, and cadmium.

1.3.1.1 Primary physicochemical parameters measured, determined, or detected

- Measures the surface tension of liquids
- Measures the viscosity of liquids
- Measures the IFT between two immiscible liquids
- Measures the wetting coefficient or wettability
- Measures the contact angle between the solid glass wall and liquid interface
- Measures the friccohesity, a dual force theory
- Measures the density up to fourth decimal place
- Determines the concentration on calibration pattern
- Measures the adsorption capacity of adsorbents: activated charcoal, silica, and alumina
- Determines the drug binding and release pattern
- Detects the length of the hydrophobic tail of series surfactants
- Detects the functional group in series
- Determines the cationic size in series of salts

1.3.1.2 Secondary physicochemical parameters measured with Borosil Mansingh Survismeter

- Measures activation energy and Gibbs free energy
- Measures entropy and tentropy
- Measures enthalpy and osmolality
- Measures the surface excess concentration
- Measures the surface area and particle size
- Measures viscosity and average molecular weights
- Measures the hydrodynamic volume
- Distinguishes between Newtonian and non-Newtonian
- Measures rheology and stress and strain
- Measures structural breaking and making effects

1.4 Distinguishing Solution and Dispersion

Expressions of molecules vis-à-vis the solvent or the medium induce interactions for a state noted as state functions. These state functions are noted as physicochemical parameters whose magnitude characterizes the nanomaterials. It becomes most significant when atoms or molecular NPs like CdS_2 and TiO_2 and atomic NPs like Au and Ag take part in such sciences. Considering all these sciences the book is designed for those users who are really involved with substantial understanding or are keen to learn the fundamentals of **physicochemical characterization as a boon for nanomaterials**. For verification of the said facts the experiments will be conducted with the survismeter. The experimental data after a statistical verification will be interpreted in the true spirit of chemical bonds, electrostatic forces, and hydrophobic and hydrophilic domains. These days physicochemical characterizations have become a backbone for material or nanomaterial formulations where an adequate measuring device is need that could track supplementary data also and then integrate the profile of information about the role of solvent, temperature, pH, pressure, surfactant, oil, and others. Since nanomaterials are seriously affected by redox activities, their measurements in a simulatory manner are tracked with the survismeter, which produces surface tension, viscosity, friccohesity, IFT, activation energy, particle size, surface area, and quantitative verification of the hydrophobic-hydrophobic activities or data that are useful for capping of NPs, detaching of particle as dirt.

- Fundamentals of PCPs developing molecular coordinate
- Brief introduction of nanomaterials
- Basis and applications of physicochemical technology
- Capping and the role of hydrophobicity
- Distinctions between solutions and dispersions on an entropic scale
- Molecular self-assemblies for better optimization
- Applications of the survismeter for academia and industries
- Nanoemulsion formulations for wettability

Chapter 2

Survismeter: A New Device

A survismeter measures surface tension, viscosity, interfacial tension (IFT), wetting coefficient, contact angle, activation energy, particle size, hydrodynamic volume, and solvent binding constant. These parameters critically define the nanoemulsion and the state of liquid mixtures. Figure 2.1 gives brief geometrical frameworks for experiments and understanding of the users. Surface tension, wetting activity, spreading ability, viscosity, activation energy, and IFT are intimately correlated with the state of the liquid mixture. At a specific temperature, pressure, volume, and structure, cohesive forces (CFs) are developed, which perform many functions if they are applied in chemical processes being extended for industrial activities. In the given figure, CPU stands for controlled pressure unit, ZEB is zero energy bands, and VFB and SFB stand for viscosity functional bulb and surface tension functional bulb, respectively. VBB is for viscosity buffer bulb and SBB is for surface tension buffer bulb.

The survismeter is a virtual valve technology (VVT); it is a thermodynamically closed system and runs on the basis of the socket-cone-joint venture (SCJV) for pressure control within a thermodynamically closed system (Fig. 2.1).

Survismeter: Fundamentals, Concepts, and Applications
Man Singh and Sunita Singh

ISBN 978-981-4774-70-3 (Hardcover), 978-0-429-02761-1 (eBook)
www.panstanford.com

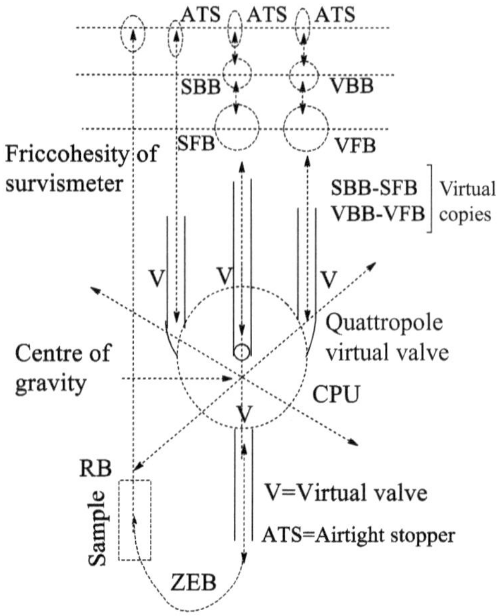

Figure 2.1 Molecular motion and pressure-driven lab device. Dynamic model of friccohesity.

2.1 Quattropolar, Virtual Valve, Center of Gravity, and CPU

The quattropole concept and virtual valves are the main levers of the survismeter that control the sample loading and measurements. The highest dimensional and spatial compatibility between the CPU and the pendant drop number (pdn) shape is evolved; otherwise the survismeter theory will fail to produce the most accurate data. The pressure in the CPU is highly optimized by adjusting the dimensional angle inside the CPU. Figure 2.2 defines that the angle above the pdn is never 90°; otherwise there would have been a pressure strike as per flux and force theory. The major theory of the CPU is that it completely prevents flux and forces from happening, which are caused by the pressure that is used for lifting the sample and allowing the sample to come back to the reservoir bulb (RB). From any side of the CPU no pressure effect is reflected in the form of the

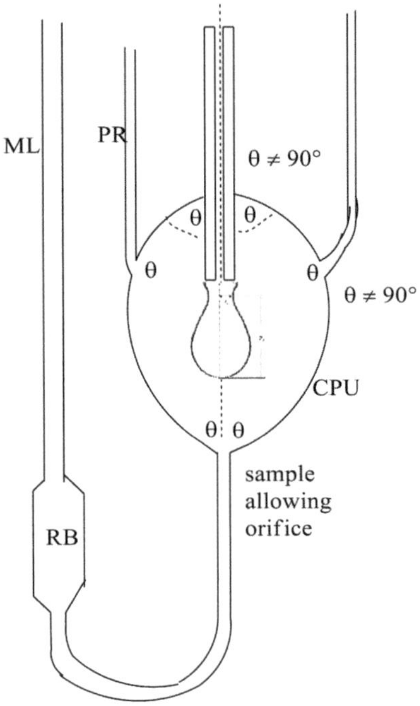

Figure 2.2 The angle above the pdn is never 90°.

flux and forces, and hence pdn of natural dimensions is formed as per the nature of the molecules present in the sample.

The VVT of thermodynamically closed systems works on the basis of the following equation:

$$\left(\frac{\partial P}{\partial V_i}\right)_{\mathrm{T,c,pH,}} = \left(\frac{\partial p}{\partial V_{\mathrm{CPU}}}\right) dV_{\mathrm{CPU}} + \left(\frac{\partial p}{\partial V_{\mathrm{SL}}}\right) dV_{\mathrm{SL}} + \left(\frac{\partial p}{\partial V_{\mathrm{FB}}}\right) dV_{\mathrm{FB}} + \left(\frac{\partial p}{\partial V_{\mathrm{BB}}}\right) dV_{\mathrm{BB}},$$

where ∂P depicts the total pressure change in the concerned operational unit and ∂p indicates the pressure changes in the CPU, SL (streamliner), FB (functional bulb), and BB (buffer bulb) units, with the $\partial V_{\mathrm{CPU}}$, ∂V_{SL}, ∂V_{FB}, and ∂V_{BB} in their air volumes replaced by the sample, respectively. Considering the partition of energy among the molecules or the relevant particles in the liquid sample quantitative rationalization could be established among the energy derived by variable coordinates.

$$\left(\eta = Ae^{-E/kT}\right) = \frac{4}{3}\pi r^2 = \frac{4}{3}\pi r^2 h = \frac{1}{2}mv^2$$

There is a possibility that the above relation could be used for developing equivalence in the energy involved in measuring viscosity, surface tension, and wetting coefficient. The survismeter measures molecular forces of solvents as well as of liquid mixtures and the structural changes that occur on solute–solvent interactions or hydration, caging, and ionic hydration.

$$E = \frac{1}{2}mv^2 \quad \text{or} \quad E = \frac{1}{2m}m^2v^2 \quad \text{or} \quad E = \frac{p^2}{2m} \quad \text{or} \quad v = \sqrt{\frac{2E}{m}}$$

The energy is determined from the Arrhenius equation, and the energy or the activation energy could be connected with the Arrhenius equation for calculation of the velocity of the molecules. The m is for mass of the molecules.

Scientifically the survismeter is an asset in the areas of nanosciences, pharmaceuticals, solution chemistry, and biochemical engineering. It saves ~95% of laboratory and human resources. There are mass fluxes and forces that help the liquid sample to move to a wanted site for distribution and energy compensation, noted as prerequisites for measurements. The weakening of CFs and strengthening of frictional forces (FFs), noted as friccohesity, predicts that the lower the surface energy and conductance, the higher is the viscosity and osmotic pressure. The survismeter device works using multiple assembling units for equilibrating natural fluid dynamics, so many localized phenomena occur, which are noted as follows.

2.2 Interzonal Distance

The interzonal distance (IZD) coordinates act as restrictions, for the following parameters:

- Kinetic energy (KE) and activation energy
- Potential energy (PE) vis-à-vis CF and intermolecular force (IMF)
- Brownian distribution of particles

- CFs of the overall mechanism
- Capping mechanism or noncapping activities
- Difference between capping and noncapping activities
- Coagulation with similar and mutual arrays
- Dispersion and IZD for optimizing
- Top and bottom engineering
- Orientational and dissipational energies
- Streaming centers and orientational sciences
- Combinatorial simulations
- Flux and forces dynamics
- Locking and unlocking physicochemical process by socket and cone interfaces (SCI)
- Pressure regulatory system
- Antspilling and antflashing

Apart from above a few more facets must be functional for successful use of the survismeter.

- Interprocess communication (IPC) connecting sample with the survismeter knobs
- Energy optimization in the BB
- Sensor as rings on glass parts or surface area
- FBs in the zone
- Streamlining zone and array developing zone
- Mode of change of guard and sample relaxation time
- Pressure regulation via pressure limb
- No pressure jerking and pressure regulatory system

2.3 Fundamental Model of Friccohesity Measurements

The basic method for measuring the friccohesity is the adhesive force (AF), which is developed on a solid glass surface and is also noted as the force working at the solid–liquid interfaces (SLIs). AF is an essential condition for measuring the friccohesity. When the liquid sample is allowed to flow within the capillary of the prescribed inner diameter (id) then the CF of the liquid sample

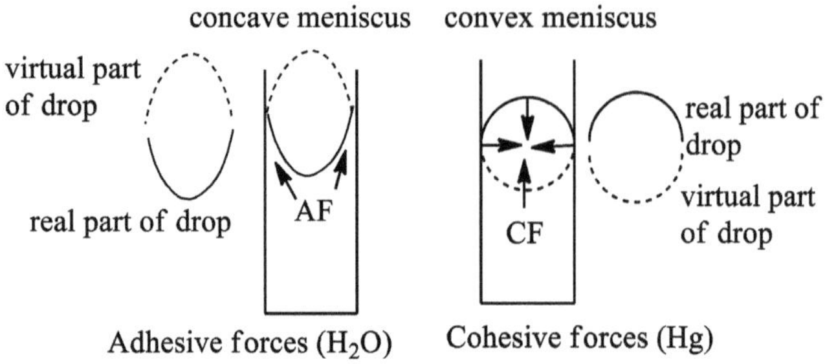

Figure 2.3 The mechanisms of AFs and CFs vis-à-vis the solid uniform capillary of the survismeter.

is transformed into an AF at the terminus of a hanging capillary of the surface tension measuring unit (STMU) of the survismeter. Figure 2.3 depicts the mechanisms of AFs and CFs vis-à-vis the solid uniform capillary of the survismeter. The dotted curvature depicts a virtual part of the pdn, whereas the solid line depicts the real part of the pdn, which is for either the concave or the convex meniscus.

Similarly since the surface tension is an expression of the unused force vectors at the air–liquid interface (ALI) or the liquid–liquid interface (LLI), which is logically depicted in Fig. 2.4, the higher the force vector at the surface, the higher is the surface tension and more intense is the down pull of the liquid at the center, which is known as the meniscus. The intensity of the pull directly reflects the nature of the molecular interactions or the activities taking place in the bulk phase of the sample. It is another proof that friccohesity is a real parameter that really explains the true picture of the pull and bulk activities of the sample in the joint venture. The pull basically attempts to develop an optimized surface that saves the maximum energy as cohesion is an antiwetting process. Such cohesion is developed because the molecules align symmetrically along the surface and at the center the maximum pull occurs. This proves that the unused forces have a larger tendency to be used by the bulk forces. In such structure-driven physicochemical expression Gibbs energy (ΔG) is

$$\Delta G = \sum_i N_i 4\pi r^2 \gamma,$$

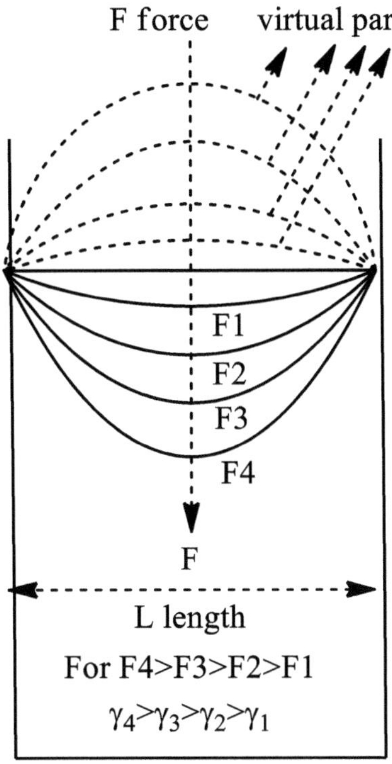

Figure 2.4 The surface forces leading to the development of a virtual drop (F: CFs or cohesion; L: capillary id; γ: surface tension of the sample).

where N_i is the total number of ith molecules in the sample, r is the id of the capillary, and γ is the surface tension of the sample. The bulk forces are completely neutralized where a change in their PE is zero ($\Delta G = 0$) but at the surface unused forces exist and hence the liquid surface experiences such a partitioning of energy due to unequal force vectors at the bulk as well as the surface. Thus a partitioning of energy and associated explanations justify an origin of friccohesity that has never been answered by any researcher; only surface tension and viscosity have always been considered to create friccohesity.

The surface forces leading to the development of a virtual drop are the forces noted in the box of Fig. 2.4. Thus, the surface forces $F1$, $F2$, $F3$, and $F4$ show great importance in developing surface tension. The higher the surface force F, the higher is the surface tension γ (Eq. 2.1), where k is a constant.

$$\gamma \propto \frac{F}{L} \quad \text{or} \quad \gamma = k\frac{F}{2L} \quad \text{or} \quad 2L\gamma = F \tag{2.1}$$

Equation 2.1 is equated to Eq. 2.2.

$$F = \eta A \frac{\partial v}{\partial y} \tag{2.2}$$

Putting the value of F from Eq. 2.1 in Eq. 2.2:

$$2L\gamma = \eta A\left(\frac{\partial v}{\partial y}\right) \quad \text{or} \quad A = \frac{2L\gamma}{\eta\left(\frac{\partial v}{\partial y}\right)} \tag{2.3}$$

In the equation, L is the id of the capillary and $\partial v/\partial y$ is the velocity gradient. In general the $\partial v/\partial y$ factor remains the same throughout the total mass transfer through a capillary of a fixed id. Thus the effective area A of the laminar layer could be determined. The hanging wire or bar or line has two surface areas in operation, and hence 2 is put with the length L. If the length is more the surface tension is less. This is because since no AFs are generated no downward pull is evolved, that is, the surface tension is less. A virtual drop is developed because the pulling force works at the surface and hence such grounds create a vacancy for simulation, which leads to the development of the concept of friccohesity. Friccohenics, or the integrated profile of CF and IMF, gives a better look of molecular activities within the liquid mixtures for understanding the interactions or processes of reactions. Mathematically if we derive a tactical correlation friccohesity is the embedded force noted as under:

$$\frac{F}{A} = \eta\frac{\partial v}{\partial y} \tag{2.4}$$

In the above equation, F/A is $\mathrm{N/m^2}$ and the velocity gradient is $\mathrm{s^{-1}}$. Incorporating these values in Eq. 2.4 a new equation, Eq. 2.5, is obtained.

$$\frac{\mathrm{N}}{\mathrm{m}^2} = \eta(\mathrm{s}^{-1}) \quad \text{or} \quad \mathrm{m}\left(\frac{\mathrm{N}}{\mathrm{m}^2}\right) = \eta(\mathrm{s}^{-1})\,\mathrm{m} \quad \text{or} \quad \mathrm{N\cdot m^{-1}} = \eta(\mathrm{m\cdot s^{-1}}) \tag{2.5}$$

The friccohesity seems to be the most functional interface of CF and IMF out of the molecular forces operating between similar and dissimilar molecules. So Eq. 2.5 is solved as Eq. 2.6, given below.

$$\begin{aligned} &\text{surface energy } (\text{J} \cdot \text{m}^{-2}) = \eta(\text{s}^{-1}) \\ \text{or} \quad &\text{surface tension } (\text{N} \cdot \text{m}^{-1}) = \eta(\text{m} \cdot \text{s}^{-1}) \end{aligned} \qquad (2.6)$$

In general, each molecule has cohesion energy, PE, or binding energy. When two molecules are brought in contact with each other then they initiate cohesion due to the similar interaction coordinates, but when two dissimilar molecules are brought together then their cohesion is affected and there is adhesion between them structurally. If there is stronger cohesion then stronger down pull is generated and stronger binding and higher surface tension occur. Thus stronger cohesion means that the molecules are interacting and are somewhere together; therefore through reengineering, molecules like cholesterol could be dissolved through adhesion. Thus friccohesity mutually resolves the issues of deposition and dispersion both by estimating their mutual activities. For example, if molecule cohesion weakens, leading to adhesion or dispersion, then both CFs and FFs become operational, which are defined by the friccohesity. The structural coordinates for generating the activities to develop cohesion are useful. For example, determining a buoyancy application for the oil samples using the upthrust force is directly related to the saturation of oil or some other structures. The pdn directly reflects the concentration and nature of the low-density liquid (LDL) or the high-density liquid (HDL). Thus the LLI science could speak about the structure, coagulation, coalescence, and dimerization sciences of molecules like dendrimer. Such sciences become useful when the molecules have void spaces and encapsulate the drug or other similar nanoparticles (NPs) therein. The patterns of encapsulations develop a unique friccohesity because the drug or the NPs that are dispersed in the medium have to come together to the void spaces with certain CFs despite the FFs that are being created by the structured solvent medium. Therefore, the fluid dynamics that are tracked with the survismeter become an inevitable need for a researcher in case of webdrimers or dendrimers [Fig. 2.5(a, b)].

(a)

(b)

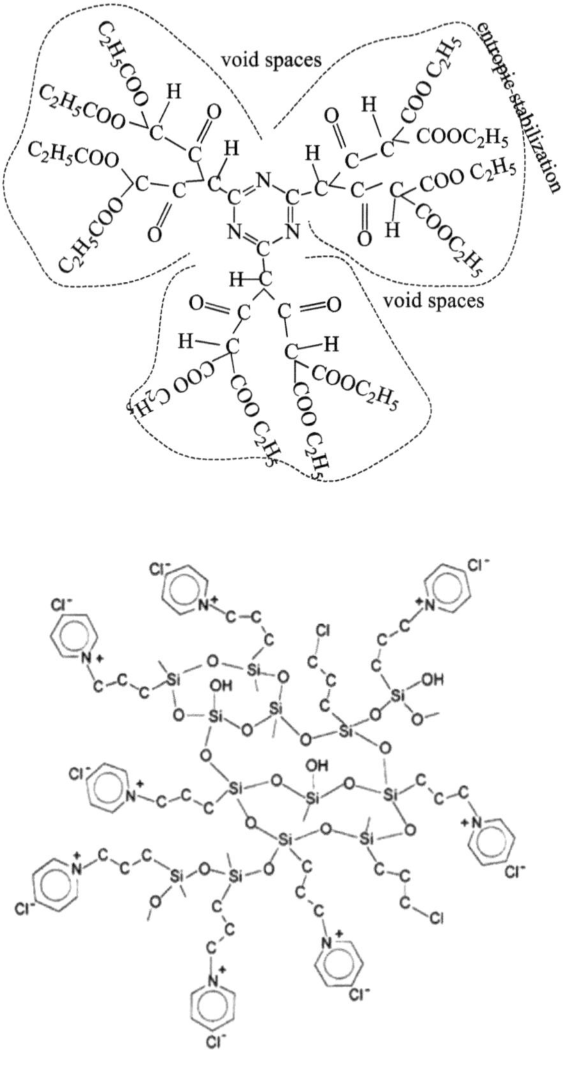

Figure 2.5 (a) 2,4,6 hexadiethylmalonate triazine: G_2 dendrimer. (b) Soluble polyelectrolyte 3(*n*-propylpyridinium) chloride silsequioxane polymer.

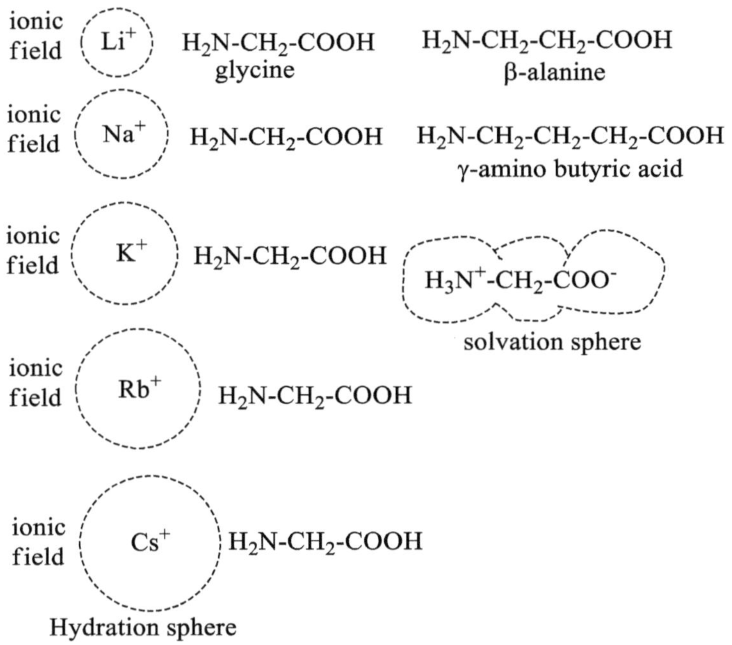

Figure 2.6 The ionic hydration and the solvation sphere of the s-block alkali salt and amino acid with an increasing number of $-CH_2$ groups, explaining their fluid dynamics and trends of friccohesity.

2.4 Advanced Method for Surface Tension Measurements

The down pull due to unused force vectors at the ALIs of the liquids is the fundamental source of surface tension. Therefore, the survismeter method exactly takes into account these expressions of the ALI, contrary to other methods, which basically involve normalization first and then measurements. The ionic hydration and the solvation sphere of the s-block alkali salt and amino acid with an increasing number of $-CH_2$ groups are depicted in Fig. 2.6 to explain the fluid dynamics and trends of the friccohesity of such systems. It ensures whether the size of the ion affects or the size of the amino acids affects the friccohesity. These studies help to understand the interaction activities of the peptides and polypeptides in variable mediums and temperatures.

Chapter 3

Electronic Optimization and Molecular Energetics

Fluid dynamics and engineering both are the foundation of sustainability, where with the selected distributions of chemical substances with required activation energy and molecular motions, they carry forward either the interacting activities or the reacting task in the chemical processes. Therefore, the desired distributions are supported by intermolecular forces (IMFs) like glucose and water. Both are mutually mixed through hydrogen bonding (HB) between them at the cost of disruption of their own HB. The cohesion of each is disrupted to allow them to interact mutually. So there could be many sets and patterns of hydrogen bonds that may exist in their mixtures; however the numbers of such sets substantially depend on their interacting stoichiometry, for example, the HB within water itself and between water and glucose and within glucose itself. And the glucose and water molecules could either release energy to the solution or could absorb energy from the solution itself. Such water-based chemical processes could transform the molecular forces of the water, but the oil is not distributed in the water because the oil has its hybridized orbital obtained by sharing of the electrons. The cohesive forces (CFs) of the oils are not easily degenerated, as is

Survismeter: Fundamentals, Concepts, and Applications
Man Singh and Sunita Singh

ISBN 978-981-4774-70-3 (Hardcover), 978-0-429-02761-1 (eBook)
www.panstanford.com

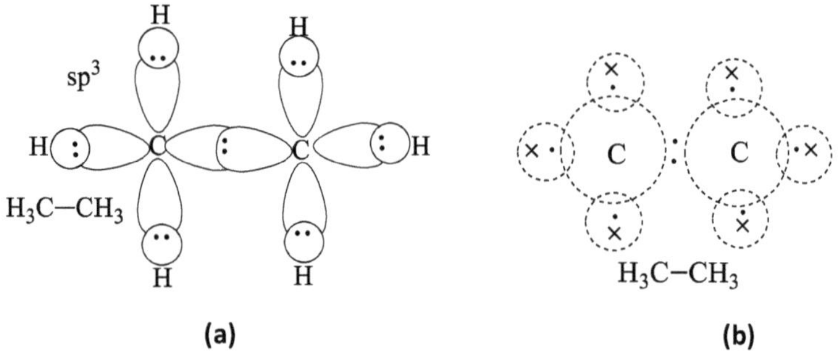

Figure 3.1 (a) Electron sharing to form sigma bonds. (b) Orbital optimizing for higher molecular stability.

explained in Fig. 3.1(a, b) (saturated), and the same is represented in the degeneration equation Eq. 3.1.

$$z = \sum_i g_i e^{-\frac{E}{RT}} \tag{3.1}$$

During interaction, the dissimilarity of electron densities leads to an activity to normalize a process and is known as the interacting activity of concerned molecules. The change in electronic energy makes molecules interact, which affects the fluid dynamics. Thus the friccohesity in terms of the velocity of the liquid flow as a fluid integrity parameter is noted in Eq. 3.2. The *e* is for electron density, the *m* is for mass of the molecules, and the *v* is for velocity of the molecules.

$$eV = \frac{1}{2}mv^2 \quad \text{or} \quad v = \sqrt{\frac{2eV}{m}} \tag{3.2}$$

The *V* is the potential caused due to a difference in electron densities of the interacting atoms in the molecules. Equation 3.2 explains the role of electrons of electronegative atoms (lowest unoccupied molecular orbital [LUMO]) for interactions and also of their sigma bonds, which have stronger hydration with the highest occupied molecular orbital (HOMO). Therefore, the saturated oils with higher potential energy and higher stability are solidified at room temperature (RT) with the least friccohesity. The shared electrons between the C–C and the C–H atoms are hybridized on the sp^3 scale with a higher potential energy and the least kinetic energy. The saturated oils with such electronic configuration tend to

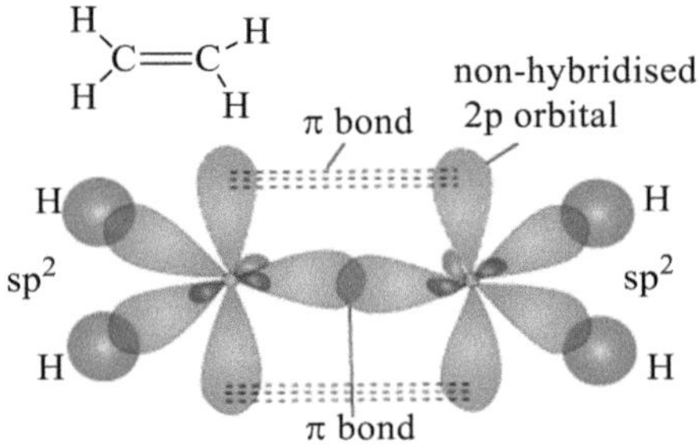

Figure 3.2 π Bonds with sp^2 hybridization of ethylene molecules.

be stabilized with stronger London dispersive forces (LDFs), which favor solidification with the least friccohesity value.

However, unsaturated oils, as is depicted in Fig. 3.2, remain liquid at RT due to the weaker CFs among their molecules or a comparatively higher friccohesity.

Therefore, the tendency of saturated fats to get solidified at RT and the unsaturated fats to remain as liquids at RT with varying viscosities could furnish valuable data about their percentage in a mixture.

3.1 Experiment: Distinguishing Saturated and Unsaturated Oil Mixtures

A survismeter can help in determining the percentage of unsaturated oil in a mixture of saturated and unsaturated oils.

3.1.1 Theory for the Experiment

Saturated oils become solid at RT (20°C) whereas unsaturated oils remain liquid at RT, so such physical states become a sensor for finding their percentages. For this purpose, initially, a standard or calibration curve is plotted between the percentage of saturated oil present in the unsaturated oil and their friccohesity and a standard empirical equation is also obtained (Fig. 3.3). The friccohesity measurements of the mixture of unsaturated oil with a fixed percentage of saturated oil are repeated at several temperatures.

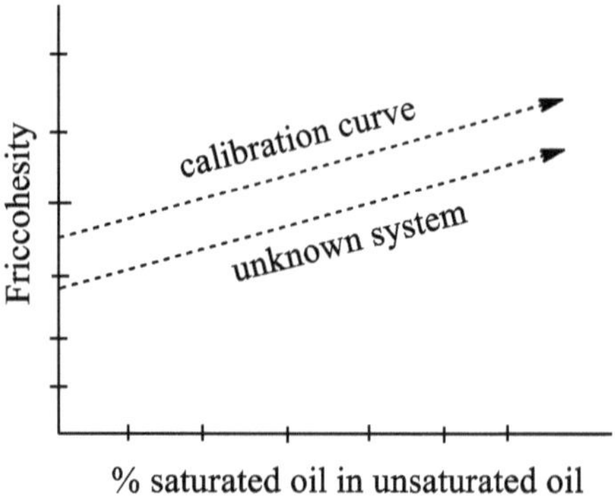

Figure 3.3 The standard calibration curve with known saturated and unsaturated oils.

The procedural details for using the survismeter can be consulted [2]. For each composition the pdn and vft data are measured with the survismeter and are plotted with friccohesity, which gives a percentage amount of saturated oil in the mixture of unsaturated oil (Fig. 3.3).

3.1.2 Systems to be Experimented

Animal fat, like that of buffalo, could be dispersed in mustard oil for determining the percentage of saturated fat present in the unknown sample. The saturated and unsaturated levels of the oils are decided by the number of single and double bonds in their structures. For example, glucose contains a hybridized orbital with a higher potential energy though the shared electron pair is shifted toward the electronegative O atom of the glucose, which makes it easily distributed in water. The two sets of molecular systems, with distribution and without distribution or with spontaneity and without spontaneity, could be developed by homogenizing them together.

3.2 Friccohesity

It is the science of molecular behavior that is especially focused on the science of HB, ionic bonding, dimerization, aggregation, and coagulations. HB depends on electrostatic forces on the molecules. So the stronger the electrostatic poles, the higher are the CFs and the lower is the friccohesity. The decrease in the surface tension of the solvent on developing a homogenous liquid mixture by adding a solute to the solvent indicates weakening of CFs of the solvent. A decrease in the CFs leads to strengthening of the HB between the solute and the solvent, or dissimilar molecules. Stronger hydrogen bonding (SHB) between the solute and the solvent increases the spreading of the solute in the solvent phase. The SHB between dissimilar molecules increases the viscosity or the movement of the solution in the solvent phase. The CFs were established between similar molecules. Addition of glycerol quantitatively disrupts the hydrogen bonds of the water molecules and diverts the water molecules to bind with themselves. So cohesion is developed by developing hydrogen bonds among similar molecules. Such interaction mechanism predicts stronger glycerol-water bonding than water–water interactions. Such molecular interaction engineering leads to a situation where the drug vehicles bind the drugs to themselves; otherwise the drugs keep dispersing in the mediums. Following this mechanism a nanoemulsion of any insoluble chemical substance could be developed in the water medium by using suitable surfactants and cosurfactants. It proves that if the drug develops a stronger binding with the medium it is difficult to release the drug to be captured by the drug vehicle for delivery in the body. Therefore, caution is maintained to avoid stronger chemical interactions of the drug with the blood. A weakening of CFs occurs only when the strength of CFs becomes weaker than the forces that develop between dissimilar molecules. Also the drug–vehicle interfacial tension (IFT) becomes operative for successful drug transportation. IFT develops micelles because similar molecules approach to develop CFs at the core as well as the periphery.

3.2.1 Friccohesity Measurement of Samples in Small Amounts: Biological and Biochemical Samples

Biological and biochemical samples are always in less quantities, so measuring their PCPs, which could also be coined as friccochemical

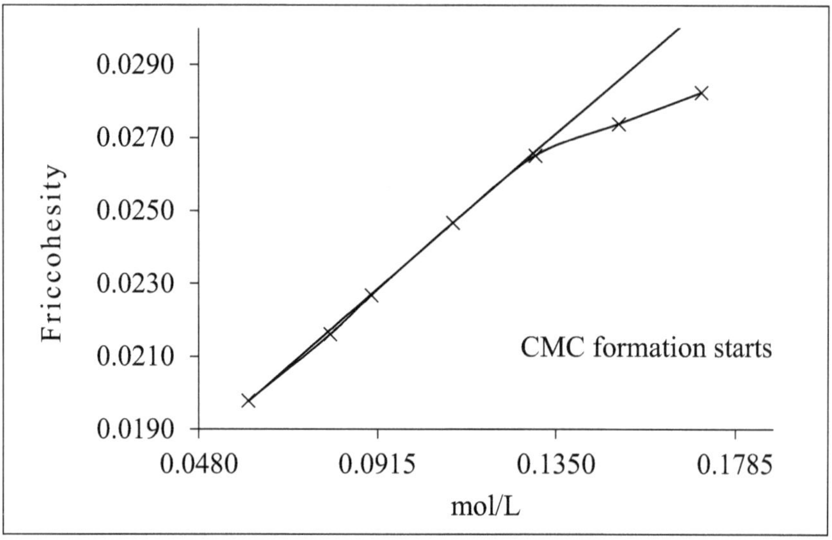

Figure 3.4 Friccohesity distinguishing critical micelles formation.

properties (FCPs), like surface tension, IFT, viscosity, surface area, particle size, and wettability, directly is not possible. FCPs are the properties generated due to an active involvement of FFs and CFs. Such samples are brought to the nanosize mode by dispersing the sample in safer solvents, like phosphate, citrate, tris buffers or tweens surfactant, and polyoxyethylene glycols (PEGs). Bringing the sample molecules to a nanosize mode ensures that their CFs are weakened and dispersed in the solvent, so the sample molecules become more functional as their binding energy is made free to catalyze their activities. When their CFs are weakened, the sample in such solvents disperse and the molecules are placed far apart from each other, generating IMFs in the infinitesimal mode, as compared to the placement of molecules in a pure state. Therefore, for measuring the viscosity, surface tension, friccohesity, IFT, surface area, particle size, and activation energy of the biological sample the survismeter is the most useful device. Most interestingly the friccohesity clearly distinguishes the micelle formation, as is noted in Fig. 3.4. The micelles are developed for optimization of the

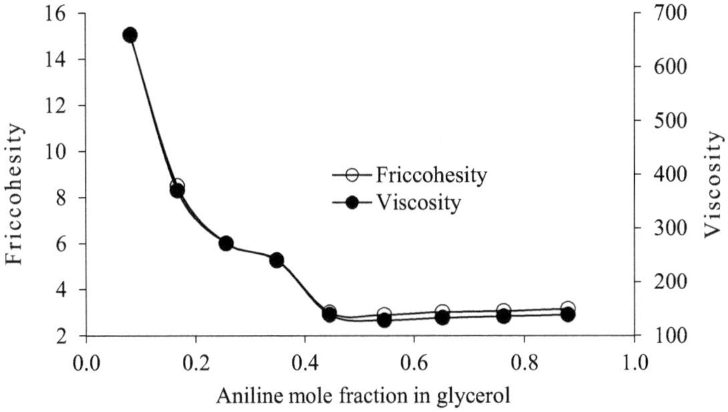

Figure 3.5 Graphs of viscosity and friccohesity.

surfactant molecules like SDS, CTAB, and tweens beyond certain their populations.

Scientifically friccohesity supersedes viscosity as viscosity alone or surface tension alone cannot resolve the molecular changes at ultramicrolevel or nanolevel. Figure 3.5 depicts the graphs of viscosity and friccohesity together. Friccohesity does not need the density values for its measurements or calculation, contrary to viscosity and surface tension.

Similarly, friccohesity also predicts the release of silibinin (SB) (Fig. 3.6), an anticancer drug, from its complex made with the TTDMM dendrimer. Therefore, the wider studies about friccohesity validate its concept and applications in solution chemistry, which could further be extended.

Similarly, survismeter measurements identify the benzene structure, which could be useful in case of identification of activities of the individual molecules with respect to the solvents and other experimental conditions. Figure 3.7 distinguishes the activities of organic molecules that differ in their structures; hence the fluid dynamics of the survismeter is so sensitive that it chemically and structurally depicts the role of the structural constitutions of molecules in specific solvents at prescribed experimental conditions.

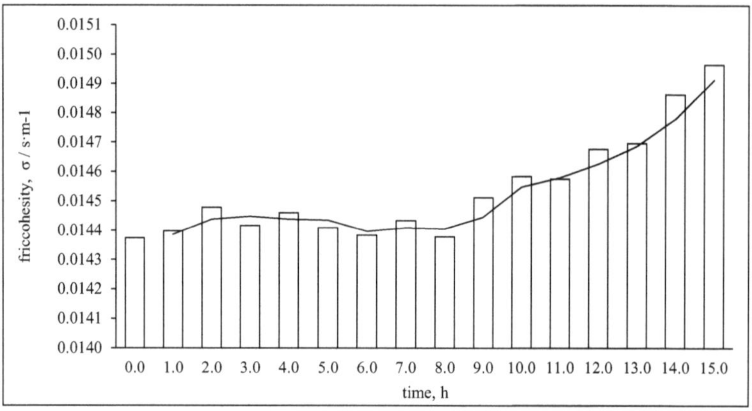

Figure 3.6 Friccohesity vs. time for SB drug release for 15 h.

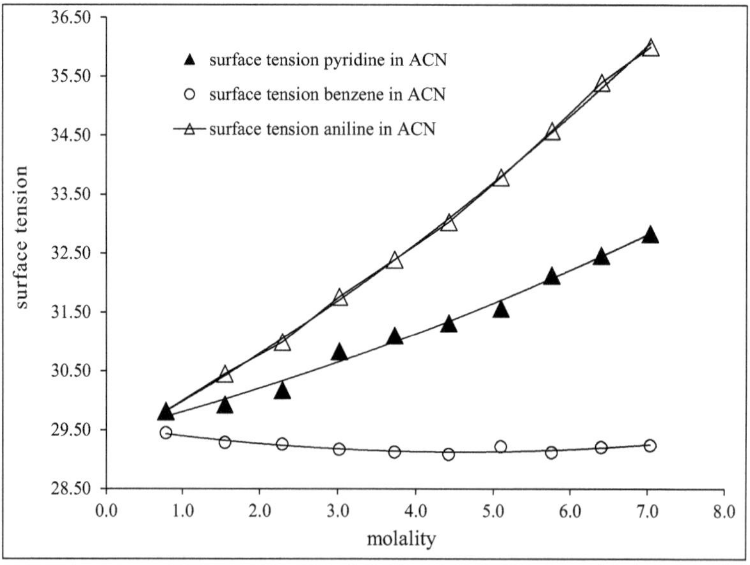

Figure 3.7 Surface tension vs. molality of organic systems.

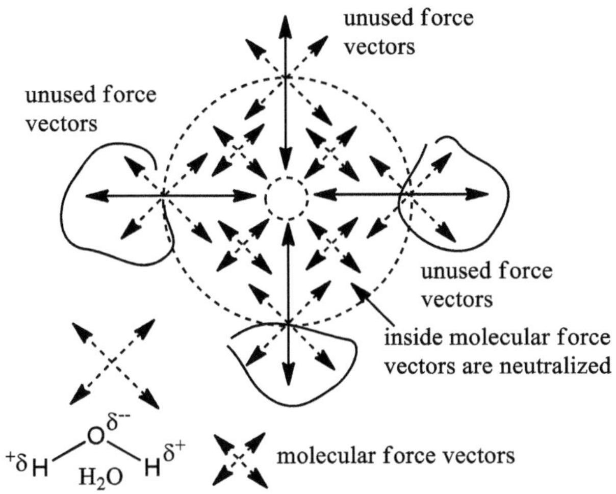

Figure 3.8 Expression of vectors of molecular forces at surface and bulk phase of the liquid sample.

3.2.2 Friccohesity to Formulate Reinforcing Materials

Friccohesity defines the fine distribution of similar particles in dissimilar particles by weakening their cohesion or CFs and leads to the establishment of stronger binding with dissimilar particles, known as interparticle binding. The particles of one species face resistance or in the process friction or frictional forces (FFs) are involved, and friccohesity deals with these two forces in a most rationalized manner. Figure 3.8 depicts that molecules at the surface tend to move toward the center and such tendencies develop cohesion, leading to the development of a waterdrop. Molecular structures with functional poles, charge accumulation on them, or permanent dipoles are suitable for forming waterdrops. Thus a similar molecular structure at the air–liquid interface (ALI) has unused forces that lead to the creation of pulling forces.

The CFs of water and benzene are not equal and develop two immiscible phases, as follows:

$$\mathrm{IFT} = \left|\mathrm{CF}_{\mathrm{Benzene}}\right| \mathrm{LLI} \left|\mathrm{CF}_{\mathrm{Water}}\right|$$

The surfactant further induces, on account of its hydrophilic and hydrophilic interactions, mutual solubilization with the solvents, noted as under:

$$\text{IFT} = \frac{\langle \text{Surfactant} \rangle}{\left|\text{CF}_{\text{Benzene}}\right| \text{LLI} \left|\text{CF}_{\text{Water}}\right|}$$

Question: Determine the hydrophobic and hydrophobic forces required to develop micelles at RT.

3.2.3 Friccohesity of Hydrophobic and Hydrophilic Forces

The following are a few examples of micelles that explain composition-driven activity for micelle formation. Figure 3.9 explains surfactant activity at its lower concentration where the surfactant molecules are scattered with weaker hydrophobic CFs but on increasing their population or the concentration, micelles are formed, aligning the philic (hydrophilic) CF and phobic (hydrophobic) CF domain together as a complete micelle. In the case of less molecular population the distance between philic and philic and also between the phobic and phobic parts is larger. At a larger distance CFs are not adequately developed and hence the alignments of the forces are not attained. The forces that work between hydrophobic alkyl chains to keep them together are known as phobic CFs. Similarly the force that works between the head and the hydrophilic part to keep them together is known as philic CF. Despite the two CFs the overall activity of the micelles works with medium molecules where these forces also tie up and equilibrate for developing certain Brownian motions. The molecular or micellar motions where the molecular arrangements keep moving in the bulk medium cause

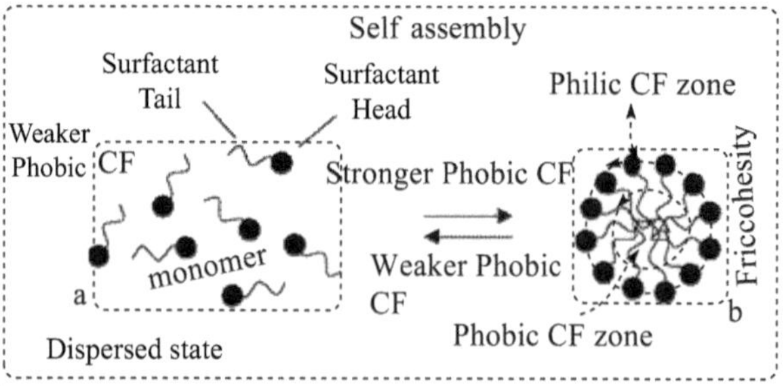

Figure 3.9 Thermodynamics and friccohesity of micelles formation.

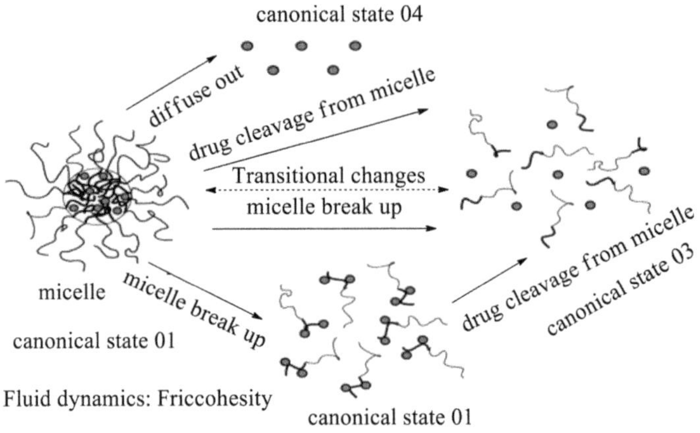

Figure 3.10 Drug cleavage from micelles.

friction. Such arrangements of forces are known as the friccohesity behavior of the micelles vis-à-vis medium molecules. The micelles have enough space, and the drugs or other entities could enter either to be detained there, for transportation, or for precipitation for the purpose of separation or release at the most appropriate site. There are two canonical boxes where the different sets of the entropy exist due to the geometrical and spatial arrangements of the monomers.

Figure 3.10 shows micelle-drug combination and how the drug diffuses out on breaking of the micelles. Similarly Fig. 3.11 distinguishes among micelles, micellar emulsion, and emulsions in terms of the coalescence of the particles being dispersed in the micellar medium. Such micellar solutions develop unique physicochemical properties or applications, especially in situations

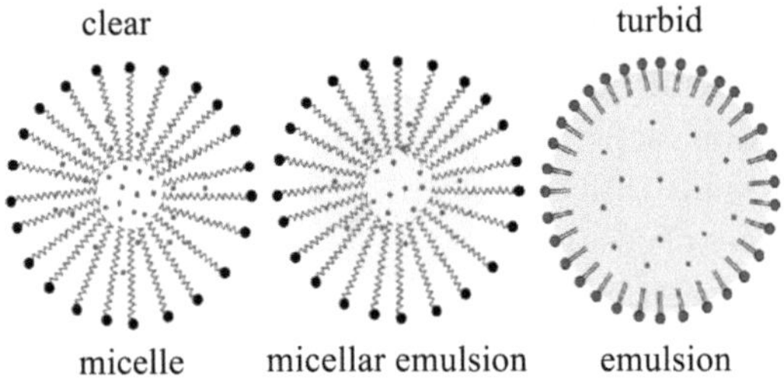

Figure 3.11 Micelles and emulsions depending on particle sizes.

where the fluid dynamics have a role to play. There are four canonical states, which have separate molecular distributions, so the systems constitute four different states of entropy production (Fig. 3.10). In such systems the medium plays a deciding role because the distribution is supported by the medium and only if the medium has a higher dielectric constant, the IMFs are stronger. Considering these features of the friccohesity activity relationship (FAR) the fluid dynamics of these samples could lead to the evolvement of a new model of sciences with wider industrial applications, like heat release or accumulation and optical properties.

The dynamics of structural activities of medium versus dispersate develop molecular forces that result in the size of the micelles as is shown in Fig. 3.11. Various kinds of micellar mixtures could be developed by either increasing or decreasing the population of the surfactants, biosurfactants, or the molecules, which could develop hydrophilic and hydrophobic forces in a most distinctive domain. The friccohesity of the micellar medium, which is depicted in Fig. 3.11, could be of interest for developing a nanoemulsion of the waste oily materials.

3.2.4 Friccohesity as a Structural Function

Friccohesity, which is a structural function of the molecular mixtures, depends on:

- Size of cations and anions vis-à-vis structures of the solvent
- Way of electronic sharing of cations and anions
- Electronegativity of anions
- Size disparity of cations and anions
- Size of surfactants with respect to their alkyl chain
- Nature of the head group of surfactants
- Structural activities of the salts, surfactants, or organic compounds with the solvent
- Activities of cosurfactants
- State of critical micellar concentration (CMC) with respect to the length of the alkyl chain
- State of loading of salts in reverse micelles and drug/dye in normal micelles

Friccohesity (σ) could be mathematically defined and designed with the following equation:

$$\sigma = Ie^{tn} \quad \text{or} \quad \ln\sigma = \ln I + tn$$

The I is for interface of the t depicted in terms of IMFs, and n is for CFs. The equation could quantitatively explain and elucidate the structural state, electrostatic linkages, HB, dimerization, and quantization of similar electrostatic charges.

3.2.5 Friccohesity to Determine Percolation of Oil or Drug Amount in Micelles

The amount of oil or drug can be determined using friccohesity measured with survismeter because the stress and strain are caused due to the central core of the micelles along with their branching size and nature during the measurements via a capillary of a fixed inner diameter (id).

The temperature effectively controls the micellar behavior. Figure 3.12 explains the temperature-dependent state of micelles. The pH also exerts effective control on micelle formation as the acid

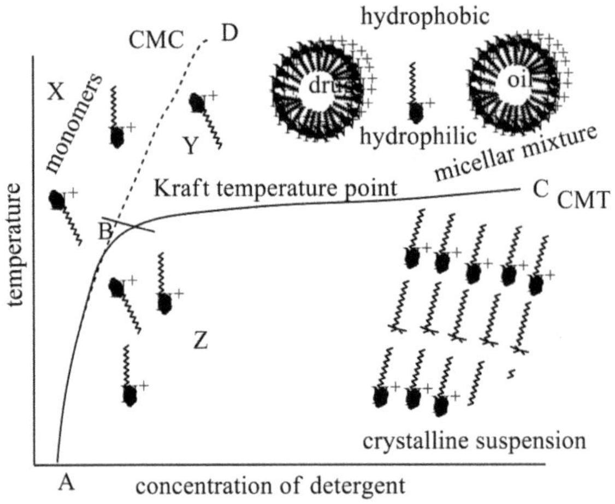

Figure 3.12 Thermodynamics of micelle science with increasing temperature and concentration of detergents.

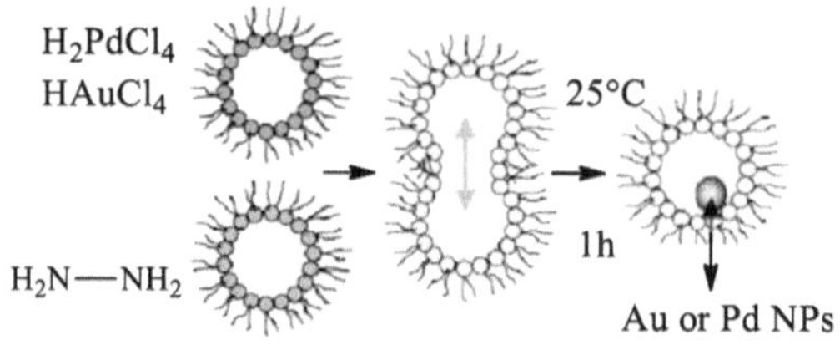

Figure 3.13 Capping of Au NPs and entering in micellar structure.

or the base contents interact with counterions of the surfactants. Therefore, the effects of the pH on micelles could be determined.

Nanoparticles (NPs) of Au and Pd can be developed via micelles, as shown in Fig. 3.13.

Micellar capping or caging type of activities of Au NPs or Pd NPs are depicted in Fig. 3.13, where the particles enter the reverse micelle, which has a hydrophilic environment.

The chemical reaction occurs as noted here:

$$2H_2PdCl_4 + N_2H_5OH \rightarrow 2Pd + 8HCl + N_2 + H_2O$$

$$4HAuCl_4 + 3N_2H_5OH \rightarrow 4Au + 16HCl + 3N_2 + 3H_2O$$

Similarly the micelles of sodium dodecyl sulphate (SDS) separate the dyes or drugs or other organic molecules. For example, SDS micelles displace the equilibrium between the crystal violet and sulfite ions in the direction of crystal violet and lead to a complete recovery of the dye. The micelles of fatty acids are effective in trapping the oily dirt particles from the soft surfaces and are used as bathing soap (Fig. 3.14), contrary to detergents, which are used as heavy-duty detergents for washing rough surfaces. The delocalization of the electronic charge of the head group further facilitates stability of the micelle because the charge distribution at the outer surface develops an effective water layer.

In the case of normal micelles the hydrophobic alkyl chain tends to be together at the center. Since the CFs are developed among the hydrophobic alkyl chain they remain together; had CFs not developed the chain would have scattered away from the center. Such science is used in developing micelles of cholesterol to wet it out from the body tissues. The monomer water may diffuse out of the hydrogen-bonded water so two phases of the solvent exist in liquid

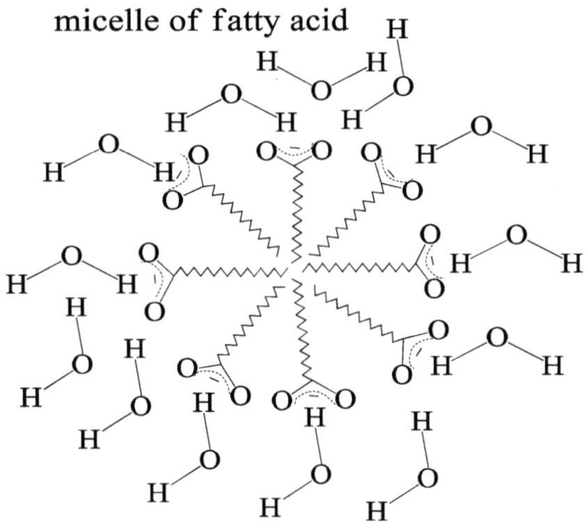

Figure 3.14 Micelles formation of fatty acids, hydrophobic alkyl chain tending toward center as core and polar water molecules aligned at periphery.

mixtures. It is similar to dendrimer and micelle formation because on adding a smaller amount of surfactant to the solvent/water the surfactant scatters or disperses haphazardly as there are enough solvent phases to accommodate dispersion but on increasing the surfactant amount micelles start to form. It proves that the larger population develops surfactant–surfactant interactions, which allow the development of stronger cohesion.

Chapter 4

Advanced Method for Surface Tension Measurements

The down pull due to unused force vectors at the ALIs of the liquids is the fundamental source of surface tension. Therefore, the survismeter method exactly takes into account these expressions of the ALI, contrary to other methods, which basically involve normalization first and then measurements. The ionic hydration and the solvation sphere of the *s*-block alkali salt and amino acid with an increasing number of $-CH_2$ groups are depicted in Fig. 4.1 to explain the fluid dynamics and trends of the friccohesity of such systems. It ensures whether the size of the ion affects or the size of the amino acids affects the friccohesity. These studies help to understand the interaction activities of the peptides and polypeptides in variable mediums and temperatures.

4.1 Limitations of du Noüy Ring Method

In this device a platinum ring is slowly lifted from the surface of a liquid. In the light of Fig. 4.2, force, F, is applied to raise the ring from the surface of the liquid and is equal to the surface tension. Fundamentally, dipping a ring in a sample acquires energy

Survismeter: Fundamentals, Concepts, and Applications
Man Singh and Sunita Singh

ISBN 978-981-4774-70-3 (Hardcover), 978-0-429-02761-1 (eBook)
www.panstanford.com

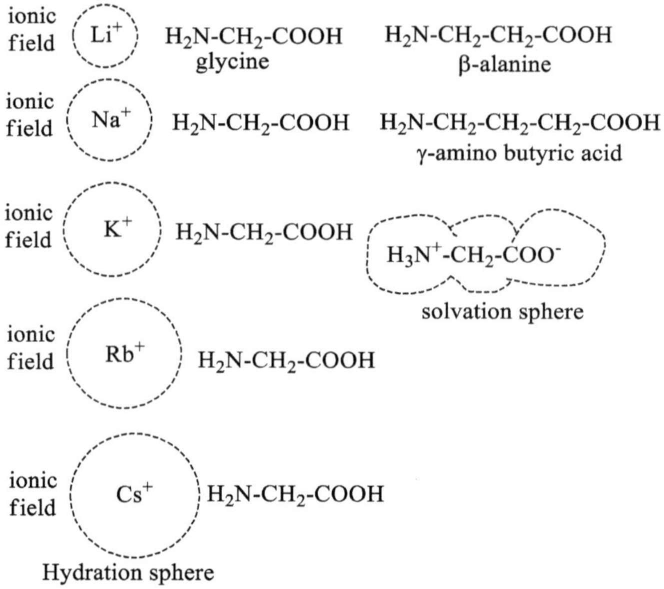

Figure 4.1 The fluid dynamics and trends of the friccohesity of ionic fields and ionic hydration.

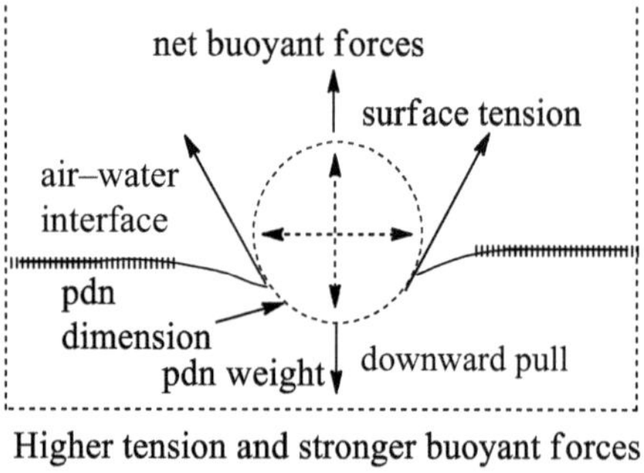

Figure 4.2 The fundamental mechanism of cohesive and adhesive forces of liquid sample that initiate development of pendant drop formation along with the solid surface of a capillary. In the survismeter, the pendant drop formation occurs in its controlled pressure unit (CPU) by avoiding pressure fluctuations and escape of sample.

and creates its own environment, so two environments exist in this method, which is against the spirit of naturally tracking and distribution of the unused surface forces of liquid samples.

Hence this method disturbs the natural state of down pull of the liquid sample where the contact angle and the geometrical simulation of the platinum (Pt) ring are never at a 90° angle. Also the air fluctuations, which are most sensitive for the stretched ring of the sample, are not controlled as the environment is open. Therefore, the formulation of forces F noted as under is wrong:

$$F = 2\pi(r_\mathrm{i} + r_\mathrm{o})\gamma \tag{4.1}$$

Here, r_i and r_o are the radii of the inner ring of liquid film pulled and of the outer ring of the liquid film, respectively. French physicist Pierre Lecomte du Noüy (1883–1947) developed this device. The survismeter device promotes expression for a more accurate utilization of the pulling force F in allowing pendant drop numbers (pdns) in an uninterrupted controlled pressure unit (CPU) environment. There is no possibility of air fluctuation to be applied on the drop size because it has many layers of protection. For example, in the survismeter there is:

- A socket at the top of the surface tension unit that controls the overhead pressure
- A top capillary just after the socket
- A buffer bulb (BB) just after capillary
- A streamlining capillary
- Functional that holds the sample
- Again a streamlining capillary

In addition, when pdns are detached the pressure is regulated with the pressure limb.

The fluctuations are controlled at the desired temperature, and a time is given for developing cohesion and adhesion for adhering of the liquid sample to the solid surface area of the hanging capillary inside the CPU, contrary to a pulling of the Pt ring. The most cutting edge of the survismeter invention is that the CPU of the survismeter has autonormalized geometry (ANG), which further facilitates an expression of surface forces at the air–liquid interface (ALI).

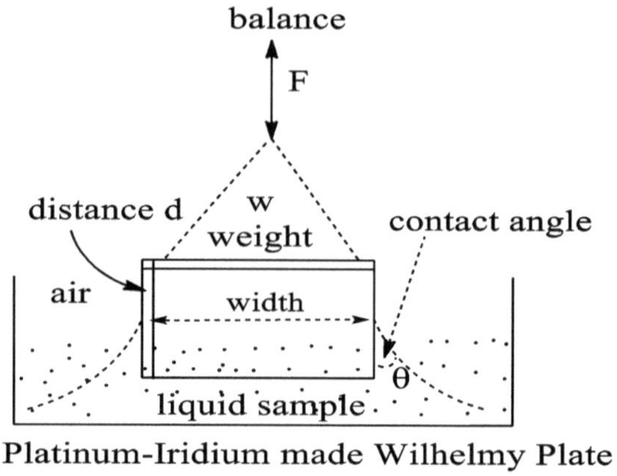

Figure 4.3 Platinum–Iridium-made Wilhelmy plate for surface tension determination.

4.2 Limitations of Platinum–Iridium-Made Wilhelmy Plate

The Wilhelmy plate (WP) is a thin plate and measures the equilibrium surface or the interfacial tension (IFT) at an ALI or a liquid–liquid interface (LLI) (Fig. 4.3). The plate is oriented perpendicular to an interface, and the force exerted on its surfaces is measured.

The WP is approximately 25 mm × 15 mm = 355 mm^2 in area. The plate is made from Pt with rough surfaces for ensuring complete wetting. However, in the experiment wetting is a great issue; sometimes air bubbles remain there; sometimes the vertical direction changes. Therefore, there is the question of smooth wetting and setting complete equilibrium between the solid and liquid surfaces. Also there is basic difference in the heat capacities of the liquid and of the plate, which is a thermodynamic consideration for measuring the surface tension. Hence, equilibrium is never attained and a contact angle of variable magnitudes could be developed. The cleaned WP is attached to a balance with a thin metal wire. The force on the plate due to wetting is measured using a tensiometer or a microbalance and used to calculate the surface tension using the Wilhelmy equation

$$\gamma = \frac{F}{l\cos\theta} \tag{4.2}$$

l is the wetted perimeter ($2w + 2d$), in which w is the plate width and d is the plate thickness, and θ is the contact angle between the liquid phase and the plate. The contact angle is rarely measured, and a complete wetting at $\theta = 0$ is assumed, which is wrong. There are many errors involved in the WP method, and again the survismeter resolves all these issue; thus the survismeter is the only way for accurate surface tension measurements as a green technology. Further there are many errors for two-phase liquid systems where the interface of a less dense phase is added on top of the undisturbed primary denser phase, which disturbs the forces operating at the LLI. Such difficulties are resolved by allowing drop formation of the high-density liquid (HDL) inside the low-density liquid (LDL) slowly, which does not disturb the forces at the LLI, by using the survismeter. Considering all deciding parameters and coinciding molecular sciences the survismeter seems to be a great breakthrough in areas of liquid mixtures, nanoemulsions, and materials sciences. Similar interfacial force factors work in the case of nanoemulsion formulations. In such formulations, the oil nanoparticles (NPs) (as hydrophobic molecules) are surrounded by the hydrophobic alkyl chain of the surfactant.

In nanoemulsions the LLI works effectively. The theory of interfacial science of nanoemulsions is evolved as hydrophobic-hydrophobic interaction (H_b^2I). Therefore, for allowing particles to spread on the solid surface H_b^2I becomes a working principle for friccohesity. Since the movement of the liquid sample for materializing the SLI there should have not been a disturbance, the CPU becomes a most relevant model to assist in friccohesity measurements.

Friccohesity directly determines the state of the ionic field that is developed on dissolving a series of electrolyte salts, like HCl, LiCl, NaCl, KCl, and RbCl, and other salts, like dendrimers from trimesoyl 1, 3, 5 tridimethyl malonate (TTDMM) to trimesoyl 1, 3, 5-tridihexyl malonate (TTDHM), in selected solvents in prescribed experimental conditions.

4.3 Friccohesity of Survismeter Measurements

Friccohesity of survismeter measurements under tactical pressure works with virtual pressure operation within the thermodynamically monitored sample reservoir. The pressure is reduced inside the survismeter ensembles, and in response to reduced pressure the liquid samples with specific cohesive forces (CFs) move upward along the uniform glass walls, causing the least friction. Both CFs as well as frictional forces (FFs) remain operational and work in effective coordination mutually during measurement of the data of selected properties. The molecules are carefully permitted back to the reservoir bulb (RB). The BB attains a similar energy distribution of the molecules after lifting the sample from it (the RB). The energy equilibrium in the RB occurs through various physicochemical hurdles, like quantum mechanics, which becomes an effective science to optimize molecular energy. Such activation energy is initiated by the molecules to streamline the flow within the structural limitations under noncovalent chemistry limits. The molecular distribution is explained via Eq. 4.3, which could go through several partition steps, noted as under:

$$\log[\eta] = \log k + a\log\left(\frac{\sum N_i M_i}{\sum N_i}\right) \tag{4.3}$$

Out of the total N molecules some molecules (N_i) get activated on acquiring activation enegry. The N_i molecules affect the interacting activities of liquid mixtures, noted as under:

$$\frac{N_i}{N} = \frac{e^{-E/kT}}{\sum_{j=1}^{M} e^{-E/kT}} \tag{4.4}$$

Thus the involvement of activation energy causes motions after disrupting the hydrogen bonding (HB), with generation of ΔH, and hence the following equation could be applied:

$$\Delta G^\circ = \Delta H^\circ - T\Delta S^\circ \tag{4.5}$$

If the liquid mixture is in equilibrium then Eq. 4.5 is modified as Eq. 4.6 because $\Delta G^\circ = 0$. Equation 4.6 depicts that the entropy could be generated either by varying the temperature or by breaking or making the chemical bonds that develop the ΔH°.

$$\Delta H^\circ = T\Delta S^\circ \tag{4.6}$$

When 1 mL of H_2SO_4 is added in 100 mL of water then what happens to the entropy of the resultant liquid mixture? Since H_2SO_4 develops exothermic reaction with water, heat is generated and the system becomes thermodynamically active. The enthalpy ΔH remains negative ($-\Delta H$), and also the pressure remains almost the same, so Eq. 4.7 is developed.

$$\Delta G^\circ = -\Delta H^\circ - T\Delta S^\circ \tag{4.7}$$

The pressure is constant, so Eq. 4.7 is obtained.

$$dq = \Delta H^\circ \quad \text{or} \quad \Delta \text{H} = T\Delta S^\circ \tag{4.8}$$

Putting $\Delta H = T\Delta S$,

$$\Delta G = -T\Delta S - T\Delta S \quad \text{or} \quad \Delta \text{G} = -2\text{T}\Delta \text{S} \tag{4.9}$$

The temperature, which generates entropy, also affects the heat contents obtained on disruption of the HB or the formation of new HB in the mixture. Figure 4.4 depicts a decrease in the surface tension and an increase in viscosity. Both the temperature as well as the composition decrease the surface tension because the force vectors are used in local oscillatory motions to dissipate the energy.

The model depicted in Fig. 4.4 defines various chemical and physicochemical activities of liquid mixtures vis-à-vis the medium or the solvents. The model is most creative, allowing fascinating interpretations and explanations incorporating structural variables for working and expression of the molecular forces, such as the antibonding model of surface tension. Such science and activities work because the molecular force factors extended toward the air phase are not fully counteracted as compared to the forces that exist with molecules in the bulk phase. Thus energywise, two phases exist in the case of a liquid sample held in the container. These forces are counterbalanced by the nearby molecular force factors, and this situation of surface forces is named the lowest unoccupied molecular orbital (LUMO) model. The LUMO model here depicts the lowest unoccupied molecular orbital. Similarly at the LLIs, surface tension is developed because at the LLI unbalanced molecular force factors are aligned and keep working, which have different electron clouds and electron density. Such constitutional molecular linings touch each other intimately and have three thin integrated layers of immiscible solvents and mutually solubilized liquid.

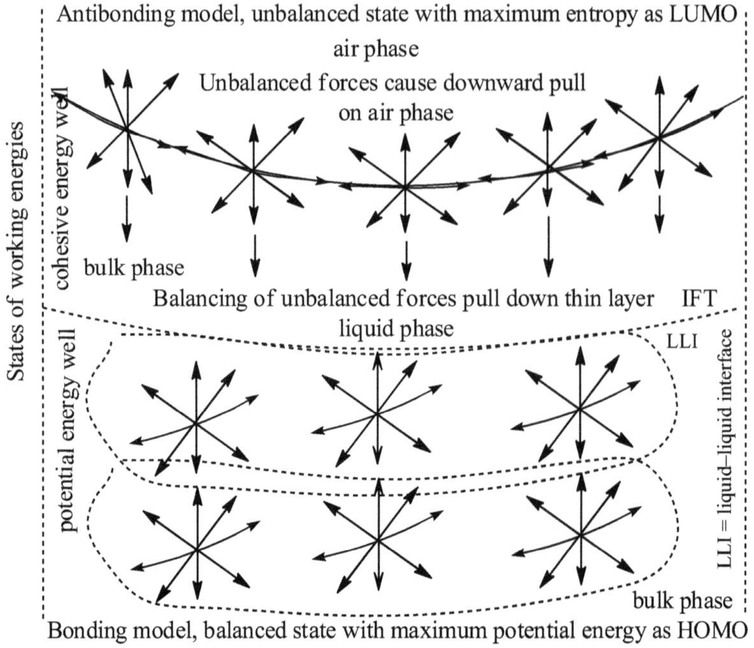

Figure 4.4 Working mechanism of IFT (interfacial tension developed due to liquid–liquid interfaces of two immiscible liquids.

However, the electronic charge and the electron cloud do not mutually share and hence both liquids, which are in close contact, do not mutually solubilize. Another reason for this science is that the hydrophobicity and hydrophilicity of the immiscible solvents also remain workable for their immiscibility and hence they need the surfactant for their mutual solubilization. Most interestingly the immiscible solvents have less contact and a higher contact angle with higher surface tension and the least friccohesity, contrary to the solvents that have a zero contact angle and mutually homogenous liquid mixtures. Mathematically, the relationship between the contact angle (θ) and friccohesity is developed through Eqs. 4.10 and 4.11. The arrows indicate the force factors that are obtained out of the integration and optimization of the electron clouds of the molecules of two immiscible solvents.

It is true that the electron clouds of one solvent are normalized and remain constant at prescribed conditions vis-à-vis another solvent. Features like arrangements of the molecular forces at the LLI get aligned in 1D geometry for better expression. However, when a nanoemulsion is developed, the surfactant alkyl chains are aligned toward the center, which holds the oil-type NPs within itself. Such geometrical mechanism of hydrophobic NPs that are centered in a hydrophobic boundary could be noted as a stoichiometric spherical alignment with definite hydrophobic forces. Cohesion among the hydrophobic and hydrophilic molecules is developed. In such arrangements London dispersive forces (LDFs) take a center stage for more friccohesive stability, which works on the basis of the intramolecular multiple force theory (IMMFT). Such model of molecular forces could quantitatively determine the amount of mutually solubilized immiscible solvents in terms of the IFT, as noted below.

$$\gamma_{ift} = \gamma_{ift}^{0} - a_i m \tag{4.10}$$

Here, γ_{ift} depicts the IFT of the LLI of the two immiscible solvents, γ_{ift}^{0} depicts its extrapolated values, a_i depicts a slope value, and m is the molality of the surfactant that is added to the LLI system. The a_i depends on the following factors:

- The nature of the immiscible solvents and the number of pi bonds of immiscible solvents
- The number of H atoms developing HB
- The number of functional groups
- The nature of electrostatic poles
- The position of functional groups as well as of electrostatic poles

Chapter 5

Friccohesity: Advanced Molecular Forces Engineering

The Greek word for friccohesity is *trivisynektikos*. The compositions decrease the surface tension because the surfactant weakens the cohesive forces (CFs) of the solvent or the medium and develops new forces with the medium. The CFs are further weakened by the temperature (Fig. 5.1).

The lower the interfacial tension (IFT), the lower is the contact angle, which indicates that the wetting is indirectly proportional to

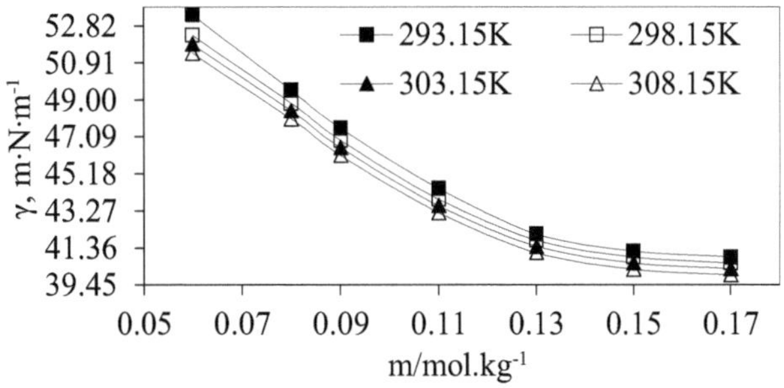

Figure 5.1 A decrease in surface tension on increasing temperature of the liquid sample.

Survismeter: Fundamentals, Concepts, and Applications
Man Singh and Sunita Singh

ISBN 978-981-4774-70-3 (Hardcover), 978-0-429-02761-1 (eBook)
www.panstanford.com

IFT values. Also the higher the friccohesity value, the lower is the contact angle, depicted in Eqs. 5.1 and 5.2.

$$\textit{Friccohesity} \propto \frac{1}{\theta} \text{ leads to } \sigma = \frac{\mathrm{k}}{\theta} \tag{5.1}$$

$$a_{\mathrm{C}} = \gamma \frac{[c]}{[c^{\mathrm{o}}]} \tag{5.2}$$

The area of the survismeter capillary is calculated using the simple mathematical relation noted as under:

$$\text{Area of the capillary} = \pi r^2 = 3.14 \times 0.5^2 = 0.785 \text{ mm}^2$$

Here, r stands for the inner diameter (id) of the capillary. The fluid flow depends on the structure of the liquid molecules and the vacant force of the solid surface of the capillary. The capillary applies opposing forces F on the unit surface area A, which creates a velocity gradient (dv/dy, cm/s/cm $= 1/\text{s}$ or s^{-1}) in fluid dynamics. The mathematical relationship and the mechanism both are explained Eq. 5.3.

$$\frac{F}{A} \propto A\left(\frac{dv}{dy}\right) \tag{5.3}$$

Eq. 5.4 is opened where η is proportionality constant, which remains constant in the case of Newtonian liquids. In the case of a liquid it does not undergo structural changes with time or with variable experimental conditions, so the viscosity remains constant and the velocity gradients also remain constant. Equation 5.4 for such arrangements is noted as under:

$$\frac{F}{A} = \eta\left(\frac{dv}{dy}\right) \tag{5.4}$$

F remains constant for the transfer of Newtonian liquids because the unit surface area (A), which remains in operation, generates a constant friction or opposing forces as the molecular arrays are in the closest of contact with each other due to adequate surface forces, which establish the force networks. Figure 5.2 explains that stronger structural development causes stronger opposing forces, which remain constant till the whole mass is transferred, so the force F acts as a rate constant.

Figure 5.3 depicts that the stronger the F, the higher is the viscosity. The viscosity keeps increasing with an increase in the composition of the systems, which depicts that the graph keeps increasing on

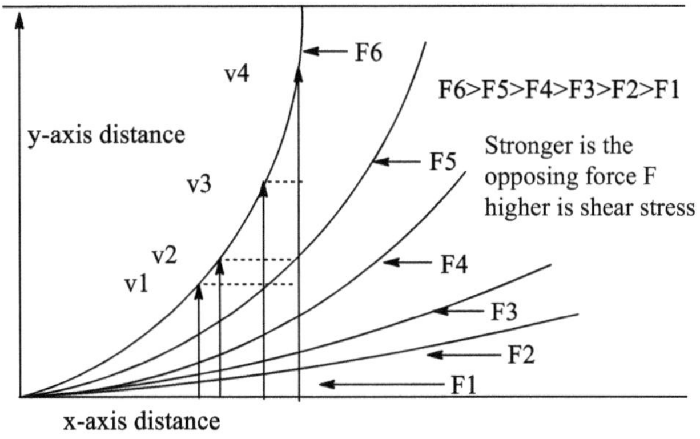

Figure 5.2 Representation of shear stress with stronger opposing forces.

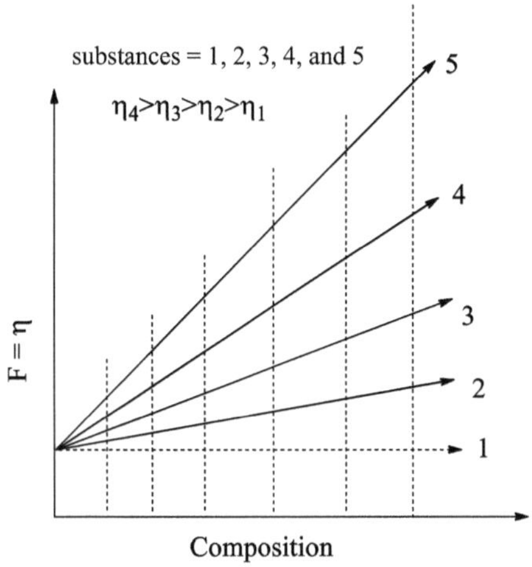

Figure 5.3 The stronger the *F*, the higher the viscosity.

the 2D scale. The liquid flows along the glass wall, and hence friction is experienced. If the same liquid is permitted to move to a larger surface area by developing adhesive forces, it forms a drop.

What happens when the solvent and the solution both are subjected to fluid dynamic measurements? How do the intermolecular forces (IMFs) develop in the liquid mixtures? How does the IMF respond to such changes? The factor that is depicted in Eq. 5.5 is the shear stress, and dv/dy is the shear strain (Eq. 5.5). So the higher the shearing causing force F, the higher is the viscosity.

$$\frac{F}{A} = \text{shear stress} \tag{5.5}$$

The shear stress depends on the opposing forces that are being applied on the unit cross section area. Since the unit surface area A plays a significant role in developing the opposing force, the capillary length and radii are important. Figure 5.2 depicts various liquids on the basis of the magnitudes of their shear stress and strain both being generated and applied on the flowing fluid, where the viscosity varies except in the case of Newtonian liquids. The linear line infers first-order processes where no activation is needed or developed due to similar mechanisms of fluid dynamics. So if the shear stress is high, the viscosity is very high.

The shear stress is a function of flexibility of the structures as well as of an interlayer binding or IMFs that are being applied on the layer adjacent to the previous layer (Figs. 5.2 and 5.4). $F1$, $F2$, $F3$, $F4$, $F5$, and $F6$ symbolize the opposing forces operational at the rigid surface of the capillary. Their sequential order is self-explanatory with the slope of the curvatures. Figure 5.2 also depicts the significance of the id of the capillary because the y-axis data are incorporated in the equation. Therefore, for developing or making a nano thin film, the distribution of nanoparticles (NPs) and fluid flow patterns are useful.

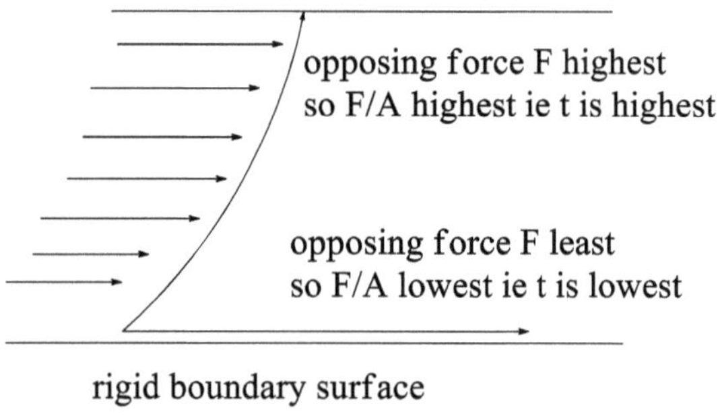

Figure 5.4 The true picture of the viscous flow with velocity gradients and the state of F being applied on laminar layers.

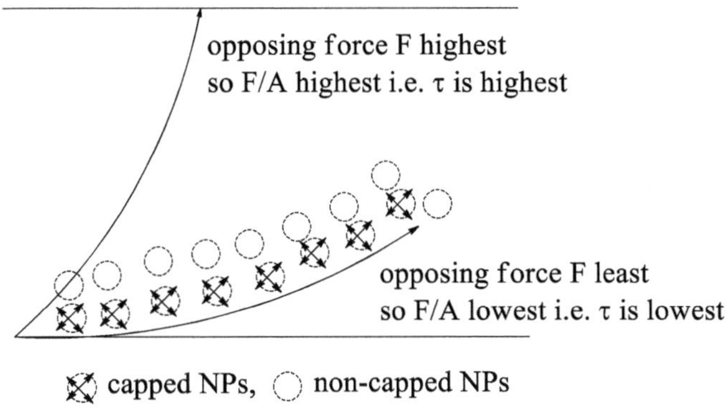

Figure 5.5 Generation of the shear with respect to the capping of the NPs like Au, Ag, TiO_2, and CdS, with variable sizes of capping agents like PVP, PVA, and alkanethiol.

Figure 5.4 illustrates the state of the shear stress and the rate of the shear strain. Such configurations are very common in the blood circulatory system, use of writing ink, syrups, nano thin film making, dyeing, and fuel cell making. The true picture of the viscous flow with velocity gradients and state of *F* being applied on laminar layers is noted in Fig. 5.4. It is like a chain of the water and the DMSO or the DMF molecules that are interlinked through the molecular network for flow.

Most significantly, Fig. 5.5 explains generation of the shear with respect to the capping of the NPs like Au, Ag, TiO_2, and CdS with variable sizes of capping agents like polyvinylpyrrolidone (PVP), polyvinyl acetate (PVA), and alkanethiol. It is clear that the survismeter could determine the mechanism and length of the alkyl chains of capping agents like alkanethiol for capping Au NPs.

In such cases truly the viscosity could elucidate the working of the binding energy and the binding stability of the agent to cap the NPs. As per Fig. 5.6 if the viscosity remains constant then the capping is of first order.

The capping itself could behave as a process to develop CFs because the NPs remain embedded in the center whereas the surface area, which has alkanethiol or another capping agent, comes in an operations process. Therefore, shear stress and shear strain could

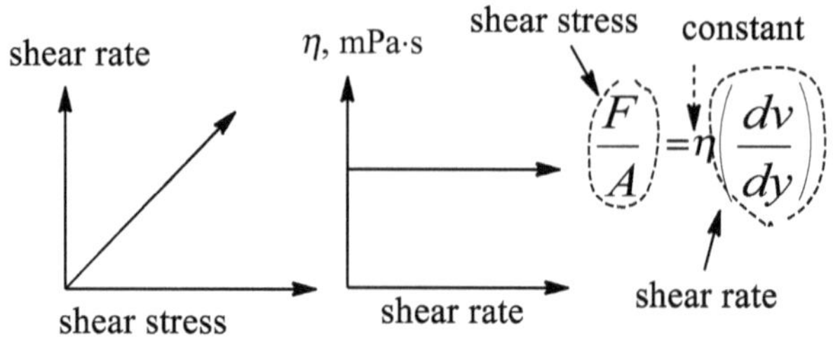

Figure 5.6 First-order capping is observed always when viscosity remains constant.

be identified with the survismeter. So viscosity and friccohesity, both which are measured with the survismeter, become the most useful experimental resources to study the science of the nanoemulsion, capping of the NPs, and drug binding and release systems of the vehicles (Fig. 5.7). Similarly, either preferential capping or competitive capping is studied by measuring the friccohesity of the NPs plus the nanoemulsion of the capping mixtures during the capping processes in definite time. Therefore, it seems that the fluid dynamics could also boost the reacting mixtures to develop fruitful collisions by accommodating the fluid flow in a specific rate flow with definite shear stress and strain of the reacting mixtures. Figure 5.7 explains that if stronger IMFs are generated due to stronger hydrogen bonding (HB) breaking or making, then the shear stress is very high.

Equations 5.6 to 5.10 explain the mathematical angles of the working liquid mixtures, which could be used as heat transferring agents, drug vehicles, or coagulating agents. Thus the relationship established among the coordinates could be explained via Eqs. 5.6 to 5.10 as under:

$$\frac{F}{A} = \tau \quad \text{and} \quad \left(\frac{dv}{dy}\right) = \text{velocity gradient} \tag{5.6}$$

Thus shear stress is a function of velocity gradient, as is noted in Eq. 5.7.

$$\tau = \eta\left(\frac{dv}{dy}\right) \tag{5.7}$$

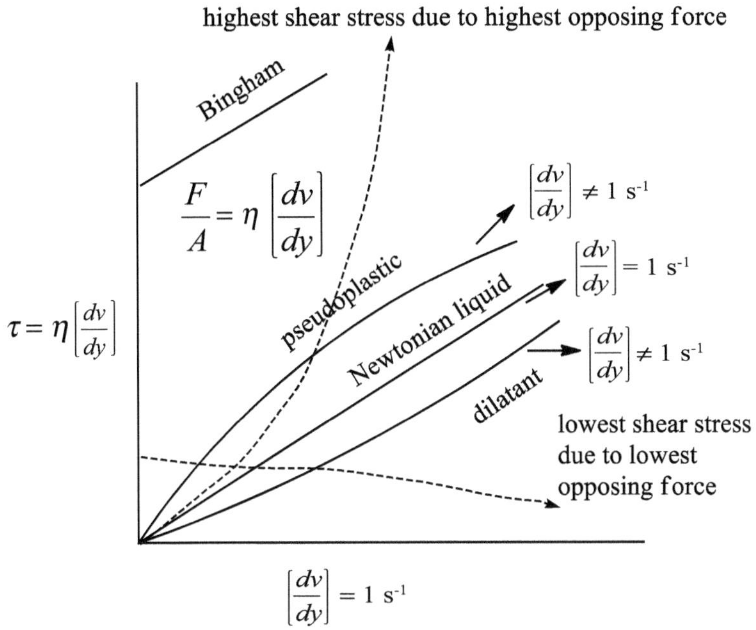

Figure 5.7 Very high shear stress observed when stronger IMFs are generated due to stronger HB breaking or making.

Or

$$\left(\frac{dv}{dy}\right) = \frac{\text{cm/ sec}}{\text{cm}} = 1\text{s}^{-1} \quad \text{so} \quad \tau = \eta \tag{5.8}$$

$$\left(\frac{dv}{dy}\right) = \frac{\text{velocity of moving layer}}{\text{interlayer distance}} = 1\text{s}^{-1} \tag{5.9}$$

$$\left(\frac{dv}{dy}\right) = 1\text{s}^{-1} \quad \text{and} \quad \left(\frac{dv}{dy}\right) \neq 1\text{s}^{-1} \tag{5.10}$$

Here, A is 1 cm^2 area, which creates an opposition or resistance to fluid flow, dv/dy is the velocity gradient, and dv is the velocity of the laminar layer at a distance y from the layer as compared to the previous stationary rigid wall of the capillary.

5.1 Science and Expression of Hydrogen Bonding

The CF is a function of HB, which is defined as is in Fig. 5.8(a–d). Figure 5.8(a) depicts the HB of the water structure, which is disrupted on adding NaCl, which develops oppositely charged hydration spheres (Fig. 5.8(b, c)). Figure 5.8(d) explains the practical situation.

Figure 5.9 explains the role of HB for mutual solubilization. The water–water molecules with similar quantum dots or electron

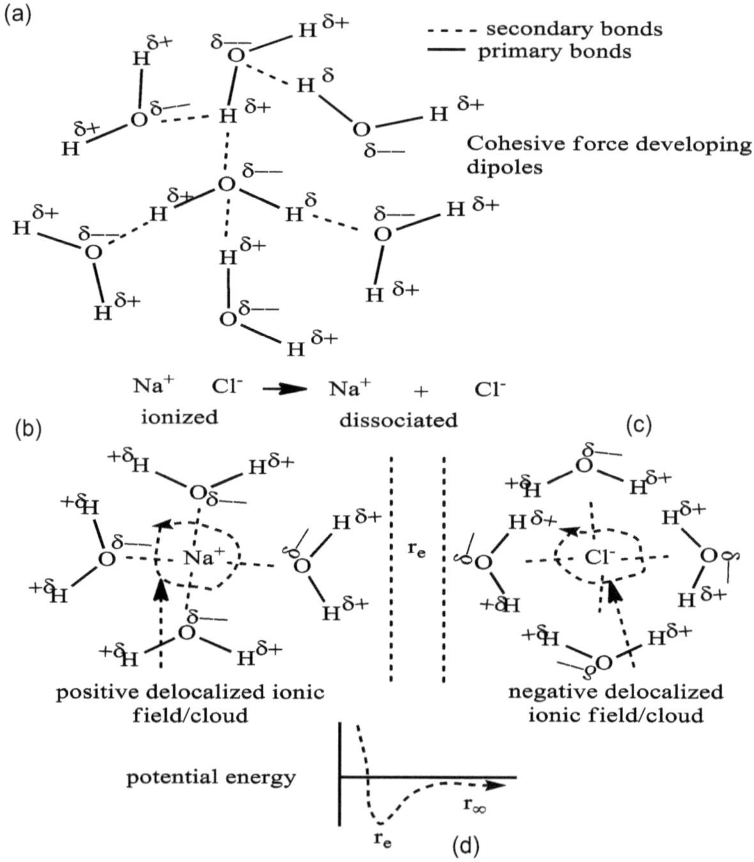

Figure 5.8 (a) Hydrogen bonded water (b) Na^+ ionic hydration (c) Cl^- ionic hydration (d) Potential energy well at equilibrium distance.

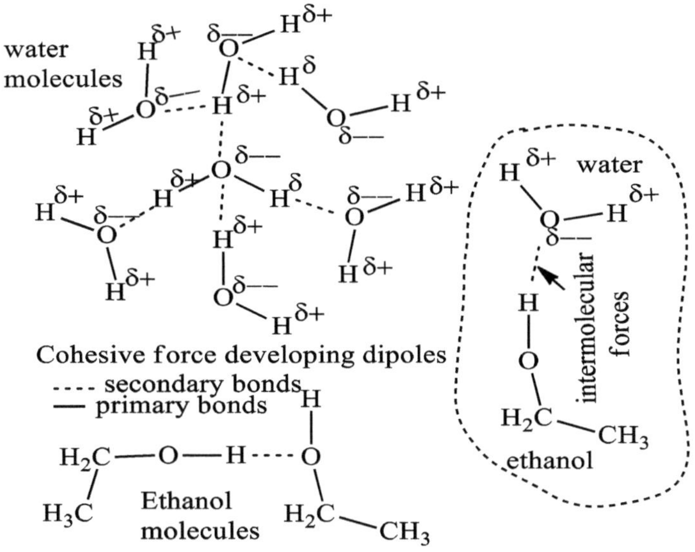

Figure 5.9 Hydrogen bonding among water, ethanol, and ethanol water molecules.

clouds with a definite resultant wave function ψ are developed for the stability of the water, contrary to the water–ethanol molecules, with dissimilar quantum dots or electron clouds with definite optimized resultant wave function (ψ). Figure 5.10 explains the development of an adhesive force in the form of a meniscus with $\theta = 0$, $\theta < 0$, and $\theta > 0$. $\theta = 0$ explains complete wetting; and $\theta < 0$ shows moderate wetting; but $\theta > 0$, with no adhesive forces, shows the weakest or almost no wetting with the solid surface.

Figure 5.11 depicts the states of the surface tension and viscosity with increasing contact angles of liquid mixtures.

The intersection of surface tension and viscosity depicts friccohesity, which explains equilibrium between both forces. Figure 5.11 has a direct correlation with the pdn formation at the lower tip of the capillary, as is depicted in Figs. 5.12 (a and b). Here, the molecular force factors that work inside the bulk pdn are counterbalanced but keep working at the outer surface of the drop and control the shape of the pdn.

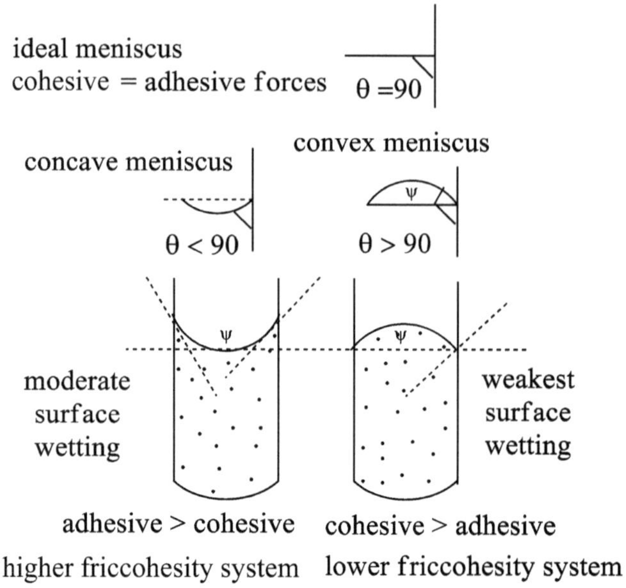

Figure 5.10 The development of an adhesive force in the form of a meniscus with $\theta = 0$, $\theta < 0$, and $\theta > 0$.

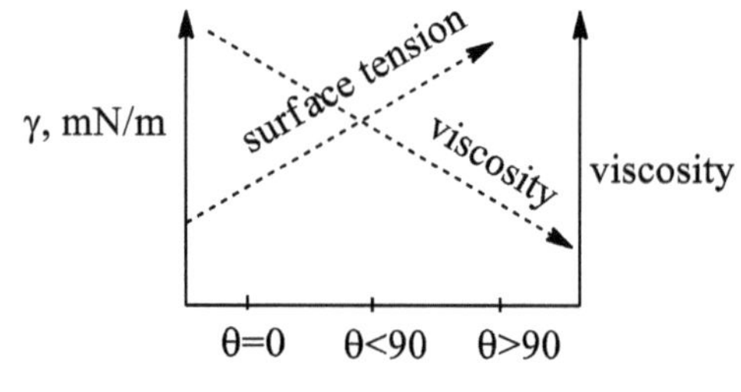

Figure 5.11 States of the surface tension and viscosity with increasing contact angles of liquid mixtures.

5.2 Scientific Capabilities of the Survismeter

The drop falls when the mass of the drop becomes equal to surface force or cohesion or tension. The contact angle is also developed with respect to the solid surface and the outer surface of the drop, as is noted in Fig. 5.12b.

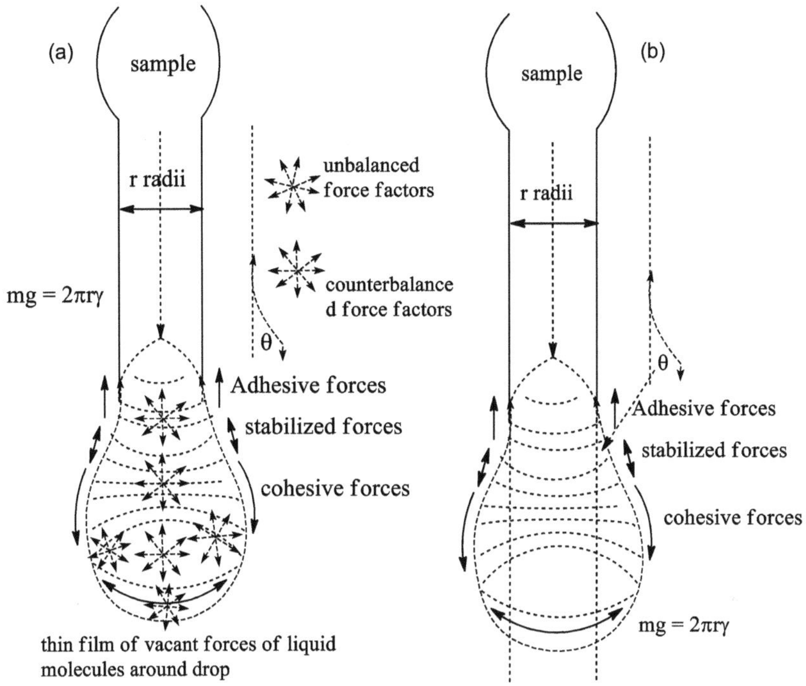

Figure 5.12 (a) Expression of adhesive forces for forming pendant drop numbers (b) Equilibrating force factors forming geometry of drop.

Considering the above-mentioned dynamics of the liquid systems the survismeter distinguishes the natural and chemical processes being conducted in liquid mixtures. For example, it distinguishes the Newtonian from the non-Newtonian liquids via Brownian motions. The measurements with the survismeter could determine an order of transformation from CFs to frictional forces (FFs), which are defined by the friccohesity. Similarly, the survismeter quantitatively determines the thermodynamic and kinetic stabilities of the liquid mixtures that are prepared for several chemical processes in which they are used as mediums, preservatives, or agrochemical formulations of pesticides and fertilizers. Such formulations induce wetting and dewetting activities, which could be experimentally determined for measuring the friccohesity. Therefore, measurements with the survismeter could be conducted for determining the following chemical activities:

- Coagulation and coalescence to purify water via friccohesity, causing higher density
- Nanoemulsion particle size determination
- Structure making or breaking in solutions and structure homogenizing in nanoemulsions
- Antisolution theory for developing single-crystal science with workable miller indices
- Antisolution activity for definite crystal planes to respond to laser or X-ray light
- Friccohesity of entropic and tentropic processes
- Free-radical scavenging by dendrimers, in general by 2,2-diphenyl-1-picrylhydrazyl (DPPH)
- Drug binding and release systems via dendrimers and magnetic nanoparticles (MNPs)
- Micelles and activated charcoal for contribution of active CH_2 groups
- Quality control of industrial experiments
- Micelles as natural scavenger or cleaner
- Low surface energy of lung fluid for good health
- Third-tier thermodynamics and phase extraction across two immiscible phases
- Dyeing, homogenizing food texture, oxygen capture, capping NPs, ionic solvation

5.3 Determining Percentage of Saturated Oil

The most significant use of the survismeter is to measure the pdn that distinguishes the percentage of saturated oils in the oil samples by lowering the temperature from 40°C to 15°C. At a lower temperature saturated oils get solidified. Thus from 40°C to 15°C the changes could be noted by determining the IFT of the water and oil phases. The oil is used as low-density liquid (LDL) phase and is taken in the RB of the survismeter. The water sample is taken in the IFT capillary. Following the usual procedure for the IFT measurements the pdn of water in oil phase is determined at variable temperatures.

5.3.1 Theory

The buoyancy application with an upthrust force is directly related to the saturation of oil or some other structure. Drop numbers directly reflect the contribution of friccohesity and the nature of the LDL or the high-density liquid (HDL). Thus the liquid-liquid interface (LLI) science could speak about the structural, coagulation, coalescence, and dimerization processes. These are chemical processes that are completed at the cost of the residual forces of the molecules where intramolecular multiple force theory (IMMFT), tentropy, and friccohesity play a significant role. The chemical processes that occur in nonenthalpic mode are effective for developing a safer and stable nanoemulsion for holding heat and allowing electrical conductance.

Chapter 6

Noncovalent Chemical Activities

In the case of liquid formulations the electronic states that cause nucleophilic and electrophilic sites and are known as electrostatic sites effectively induce several chemical activities, like van der Waals forces, London dispersion forces (LDFs), and hydrogen bonding (HB). In such activities, the strength of and distance between the HB is the most viable root to establish links with interacting molecules. Therefore, for formulating liquid mixtures the study and identification of HB sites is helpful. For example, the distance between the HB of water and water and water and ethanol are not the same, which develop separate physicochemical properties. Therefore, the Lennard-Jones potential (LJP) effectively works in their formulation sciences. Such a situation leads to the development of an effective friccohesity domain because the water and water could develop a definite cohesive force (CF) with definite optimization or the distribution of the force vectors. Also the water and ethanol both develop different CFs with definite molecular interacting abilities. So it is essential to consider the approximate LJP interaction between a pair of neutral atoms or molecules. In the year 1924, John Lennard-Jones initially proposed an interatomic

Survismeter: Fundamentals, Concepts, and Applications
Man Singh and Sunita Singh

ISBN 978-981-4774-70-3 (Hardcover), 978-0-429-02761-1 (eBook)
www.panstanford.com

potential model noted as

$$V_{\mathrm{LJ}} = 4\varepsilon\left[\left(\frac{\sigma}{r}\right)^{12} - \left(\frac{\sigma}{r}\right)^{6}\right] \quad \text{or} \quad \varepsilon\left[\left(\frac{r_{\mathrm{m}}}{r}\right)^{12} - \left(\frac{r_{\mathrm{m}}}{r}\right)^{6}\right] \tag{6.1}$$

Here, ε is the depth of the potential well, σ is a well-defined distance where the interparticle potential is zero, and r is the distance between the particles. Similarly r_{m} is the distance at which the potential reaches its minimum. At r_{m}, the potential function has a negative ε value. The distances are related as $r_{\mathrm{m}} = 2^{1/6}$ and $\sigma \approx 1.122$, and these parameters are fitted to reproduce experimental data or quantum chemistry calculations. The LJP is widely used in computer simulation.

6.1 Density Determination with a Survismeter

The density could be determined with a survismeter using the following equations:

$$\rho \propto \left(\frac{dv}{dt}\right)_{\mathrm{T,\ P,\ c}} \quad \text{or} \quad \rho = k\left(\frac{dv}{dt}\right) \tag{6.2}$$

Equation 6.2 could be extended to the new equation, Eq. 6.3, noted as under:

$$\rho = \left[\left(\frac{\sigma}{\sigma_0}\right)\left(\frac{n}{n_0}\right)\left(\frac{t}{t_0}\right)\right]\rho_0 \tag{6.3}$$

Here, ρ and ρ_0 are densities of sample and solvent, respectively; dv is the total volume in the functional bulb (FB); n and n_0 are pendant drop number (pdn) for sample and solvent, respectively; t and t_0 are viscous flow times for sample and solvent, respectively; and σ and σ_0 are friccohesities for sample and solvent, respectively. Currently physicochemical data have been indispensable in basic as well as interdisciplinary, industrial, and academic understanding of chemical processes. Therefore, the electrostriction or electrostatic poles, functional groups, ionic sizes, and electronic clouds of atoms in molecules and their geometrical symmetry generate physicochemical properties that act as an authentic database to depict the working mechanism of liquid mixtures.

$$X = f[(x_{\text{electrostriction}} + x_{\text{functional group}} + x_{\text{symmetric element}})] \tag{6.4}$$

Here, X depicts a physicochemical property, like density or friccohesity and x with a subscript depicts the individual contribution of the denoted factors toward X data. Such interdisciplinaries are emerging as the thrust areas of current academic as well as industrial applications. Combinations and Boltzmann energy distribution via LDFs on the LJP lead to the development of stable thermodynamic states of the resultant liquid mixture via $e^{-E/kT}$. Therefore, structural interactions with physicochemical data could be correctly defined and explained for developing advanced and new models. So there is an urgent need to develop a homogenous solution that supports the single mechanism of all the molecules in one energy state by performing activities and partition functions for establishing a single orientation, single activation, and single molecular activities in the whole compartment. Therefore, the activities of solute are synchronized vis-à-vis their networking with the activities of solvent, which turn out to be the most accurate operational approach of the survismeter. Such interfaces of activities lead to the production of accurate solutions for developing ability and identification of nanoemulsions. The solute-solvent synchronizations work within several compartments of the survismeter with respect to the rigid geometrical boundaries of the survismeter, which leads to the development of the functional arrays out of partitioning or distributing theories. Such a study furnishes valuable information about the chemical processes involving a chemical mechanism with or without in-situ transitions. Therefore, resolving and identifying chemical mechanism in a process is explained by applying a tool like the survismeter, which accurately resolves the chemical processes. For a single composition, temperature, orientation, and canonical patches cause quantization, similar to light quanta, noted as $E = h\nu$, where h is Planck's constant and ν is the frequency of the light wave. The density could be determined using pdn formed from the used volume and the CFs (surface tension) measured with a survismeter. The mathematical relationship is noted as under:

$$n_0 \propto V_0 \quad \text{or} \quad n_0 = \rho_0 V_0 \tag{6.5}$$

$$n \propto V \quad \text{or} \quad n = \rho V \tag{6.6}$$

Here, n_0 and n are pdn for solvent and solution, respectively, and V_0 and V are volumes filled in the FB of the survismeter. Since the volume of the FB is constant $V_0 = V$. Putting the V_0 value from Eq. 6.5 into Eq. 6.8:

$$\frac{n_0}{\rho_0} = V_0 \tag{6.7}$$

and

$$\frac{n_0}{\rho_0} = \frac{n}{\rho} \quad \text{or} \quad \rho = \left(\frac{n}{n_0}\right)\rho_0 \tag{6.8}$$

The drop numbers of the solvent and the solutions are measured with the survismeter, and the solvent density is taken from the literature data. The survismeter is useful for determining several physicochemical properties in order to study the following features for a better understanding of the interaction mechanism.

- Structural activations and reorientation of the molecules
- HB breaking and making activities
- Structural changes in the HB
- Molecular motions and van der Waals forces
- LJP and Coulombic forces
- LDFs and intramolecular multiple theory (IMMFT) and tentropy study
- New hydrogen bond formation and motion–motion interaction
- Protein docking, separation of CF patches, distribution, coagulation, and coalescence

6.2 Alignment of Cations

The surfactant has an increasing alkyl chain, which acts as a tail and is involved in hydrophobic interactions. Such interacting mechanisms lead to the alignment of the heads and tails of the surfactants during chemical processes. In such cases the friccohesity acts as significant transport data as it integrates all the activities of the tail and head in a most symmetric manner. The nanoemulsion is a wide area of research with huge applications, and there is a need to bring them to the notice and also develop some new applications, especially for the biocompatible nanoemulsion.

6.3 Biocompatible Nanoemulsion

Phospholipids and lipoproteins with suitable stoichiometric ratios and in selected solvents could develop the most stable nanoemulsions. For example, lecithin-based nanoemulsions for aceclofenac (ACF) skin delivery could be formulated by utilizing sucrose palmitate (SP) and sucrose stearate (SS) esters as the stabilizers and penetration enhancers for their thermodynamic and kinetic stability. In general, definite boundaries and equal distribution of hydrophobic and hydrophobic nanoparticles in surfactant mixtures with the best homogenization could produce a drug-loaded nanoemulsion with the smallest possible droplet size and a narrow size distribution. This type of approach is noted as a combined mixture-process experimental design. The structures of both SP and SS esters, along with sucrose diclofenac, assist in nanoemulsion formulations; their structures are noted in Fig. 6.1.

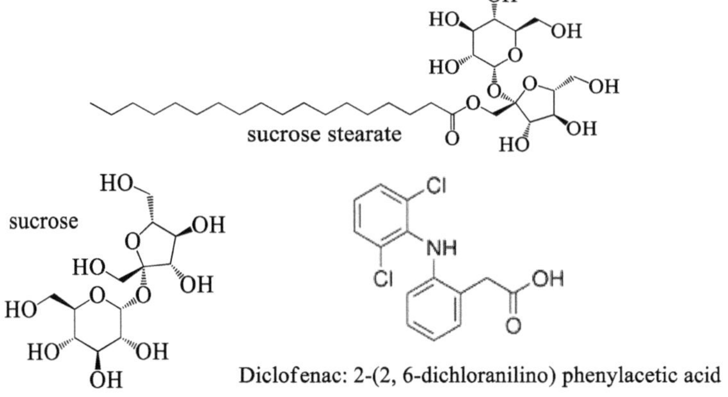

Figure 6.1 Structures of both SP and SS esters and sucrose diclofenac.

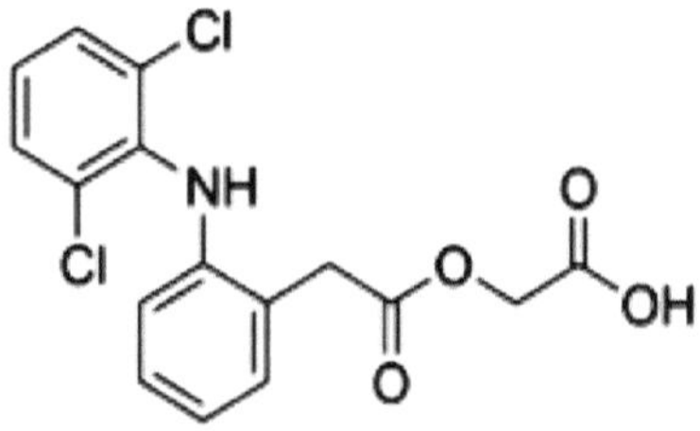

Figure 6.2 Structure of 2-(2′,6′-dichloro-4′-hydroxy-phenylamino) phenyl acetoxyacetic acid.

These molecular moieties develop the most biocompatible nanoemulsions, which could also immobilize bacterial growth. ACF (354.19 g mol^{-1}) is a crystalline powder and is insoluble in water, with higher permeability, which makes it metabolize in human hepatocytes and human microsomes, forming [2-(2′,6′-dichloro-4′-hydroxy-phenylamino) phenyl] acetoxyacetic acid as a major metabolite. The biopharmaceutical systems in drug substances depend on solubility and permeability, where ACF is a poorly soluble and highly permeable drug. These are the reasons that nanoemulsion formation in aqueous mixtures is needed for better intake. Therefore, optimization data regarding morphology, surface charge, drug-excipient interactions, physical stability, and in vivo skin performances (penetration and irritation potential) of nanoemulsions could be evaluated. The physicochemical properties and storage stability satisfy ACF-loaded nanoemulsions containing 2% of SP in a blend with 0%–1% of SS and 1%–2% of egg lecithin (produced at 50°C/20 cycles/800 bar). The in vivo tape stripping has superior ACF skin absorption from these nanoemulsions, containing 2% of SP, 0.5% of SS, and 1.5% of egg lecithin, as compared with a sample costabilized by a conventional nonionic surfactant namely the Tween 80 could of interest. Hence for a combined mixture-process the experimental design is shown as a feasible tool for development of formulations of multisurfactant-based nanosized delivery systems with potentially improved overall product performances. Thereby, kinetic and potential energies at specific temperatures or composition that make nanoemulsions stable are explained in Figs. 6.3 and 6.4.

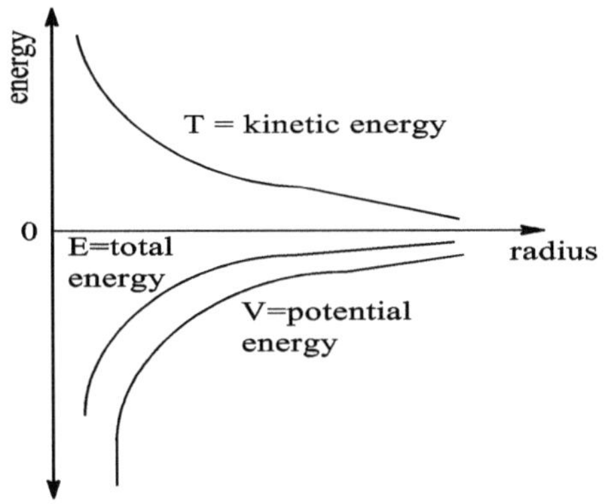

Figure 6.3 At a higher temperature potential energy is transformed into kinetic energy.

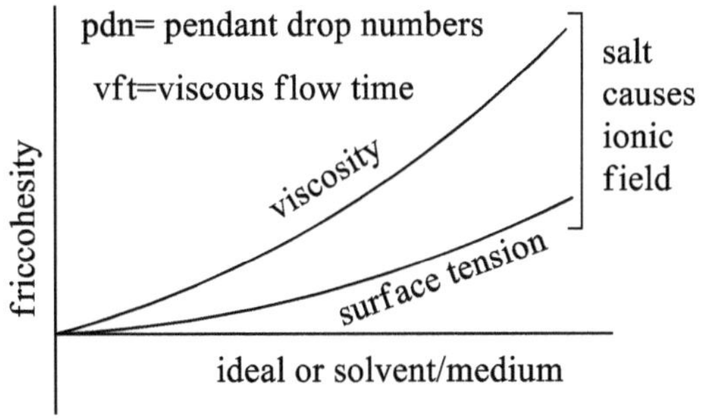

Figure 6.4 Friccohesity of aqueous salt solutions and its relationship with viscosity and surface tension.

Figure 6.3 explains that at a higher temperature potential energy is transformed into kinetic energy. Figure 6.4 finds that if both pdn and viscous flow time (vft) increase then the sample belongs to a category of ionic mixtures in which the cations and anions develop oppositely charged hydration spheres with stronger CFs; so both surface tension and viscosity increase in the case of such samples. Such interactions increase the friccohesity slightly, which is explained in Eq. 6.9, noted as under:

$$\sigma \propto (t \cdot n) \tag{6.9}$$

Chapter 7

Hyperconjugation and Friccohesity

Effects like hyperconjugation (Fig. 7.1), with definite negative and positive poles, become most prominent, inducing nanoemulsion-developing activities with definite friccohesity data. Such data could act as a sensor for the identification of organic molecules.

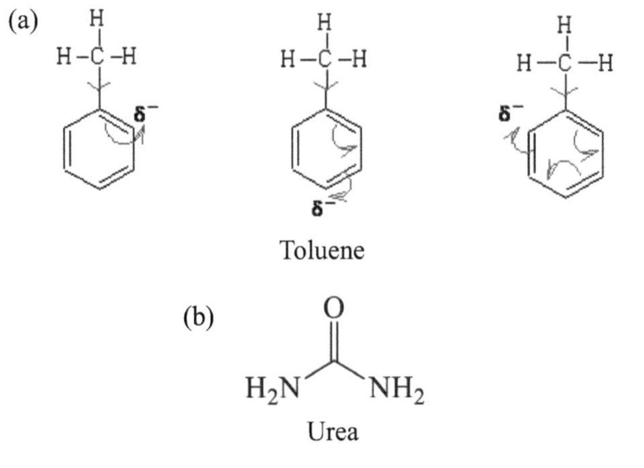

Figure 7.1 Hyperconjugation: Sigma electron delocalization (a) Hyperconjugation (sigma electron localization) in Toluene. (b) Structure of urea.

Survismeter: Fundamentals, Concepts, and Applications
Man Singh and Sunita Singh

ISBN 978-981-4774-70-3 (Hardcover), 978-0-429-02761-1 (eBook)
www.panstanford.com

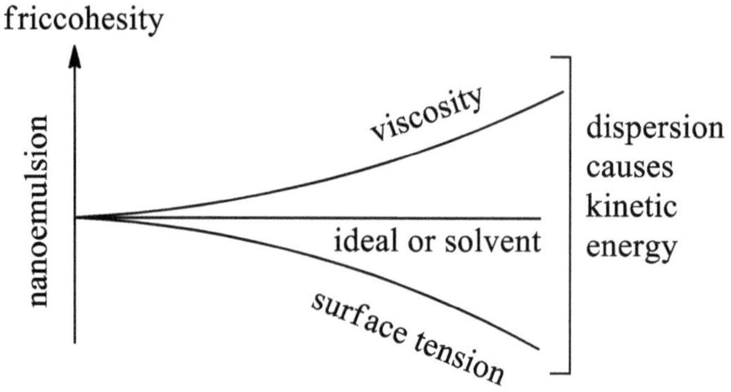

Figure 7.2 Relationship between the viscosity and surface tension in regard to the nanoemulsions.

Therefore, in the case of nanoemulsions the viscosity increases and the surface tension decreases because the cohesive forces (CFs) of the solvent are utilized to work for dispersing the surfactant, protein, or drug particles. Thereby, the effect of hyperconjugation on such activities with these systems becomes a matter of advanced research. Such changes with variables like composition, temperature, and pH could be plotted on the x-axis, as is seen in Fig. 7.2. Therefore, Gibbs energy could be calculated in terms of the friccohesity or the decreasing surface tension as well.

7.1 Anticoagulants

The molecules or drugs that prevent blood coagulation or elongate the clotting time naturally occur in leeches and blood-sucking insects. The pharmaceuticals noted as anticoagulants are used as medication for thrombotic disorders. A few anticoagulants are used in medical equipment, like test tubes, blood transfusion bags, and renal dialysis equipment. Anticoagulants are related to antiplatelet drugs and thrombolytic drugs that manipulate various pathways of blood coagulation. The development of finer nanoemulsions of these systems could be most efficient and widely useful. Thus, anticoagulants basically manipulate the coagulation cascade that builds upon the initial platelet thrombus. Such biochemical activities develop an excellent theme for research work, especially the thermodynamics and kinetic stability of such colloids or the nanoemulsion at variable pH and temperatures. Since viscosity and surface tension data, which are synergized through friccohesity, directly reflect transport and distribution activities, their measurements for such systems could furnish valuable information, which could also offer solutions to various chemical activities, like coagulation, anticoagulation, and coalescence. Currently, the biggest risk of anticoagulation therapy has been the increased risk of bleeding. In healthy people, the increased risk of bleeding is minimal, but those who have had recent surgery, cerebral aneurysms, and other conditions may have a great risk of bleeding. Fundamentally, the high level of thermodynamic and kinetic stabilities of the nanoemulsions of drugs could offer one of the best remedies because of the fact that they have the best percolation and thin film forming capacity during use. The indications for anticoagulant therapy include the following:

- Atrial fibrillation forms an atrial appendage clot.
- Deep vein thrombosis leads to pulmonary embolism and ischemic stroke.
- Hypercoagulable states (Factor V Leiden) can lead to deep vein thrombosis.
- Myocardial infarction, pulmonary embolism, and restenosis can occur from stents.
- Anticoagulation therapy prevents growth of clots.

Anticoagulation involves noninvasive pretest stratifications due to the potential for bleeds while on blood-thinning agents. Some foods and food supplements with blood-thinning effects are nattokinase, lumbrokinase, beer, bilberry, celery, cranberries, fish oil, garlic, ginger, ginkgo, ginseng, green tea, horse chestnut, licorice, niacin, onion, papaya, pomegranate, red clover, soybean, turmeric, wheatgrass, and willow bark. Many herbal supplements have blood-thinning properties, such as danshen and feverfew, and multivitamins that do not interact with clotting could be used for patients on anticoagulants. Foods and supplements like alfalfa, avocado, cat's claw, coenzyme, and spinach encourage clotting. Their intake is avoided while taking anticoagulants or if coagulability is monitored. Grapefruit interferes with some anticoagulant drugs,

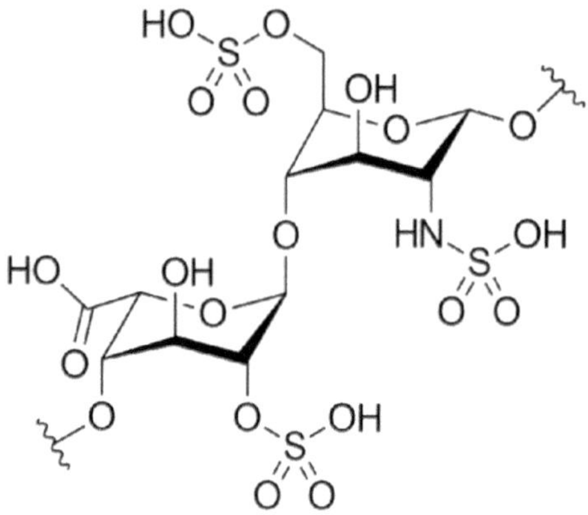

Figure 7.3 Structure of heparin molecule.

increasing the amount of time it takes for them to be metabolized out of the body. Anticoagulants are used to treat acute deep-vein thrombosis, and their activities in variable stoichiometric ratios could be studied for anticoagulation activities with the survismeter and the oscosurvismeter.

Anticoagulants such as warfarin, coumarins, and heparins are widely used. Also agents such as dabigatran, rivaroxaban, apixaban, and edoxaban are better than the coumarins, with less serious side effects. Vitamin K as an antagonist is an oral anticoagulant like warfarin and takes 48 h $\approx$ 72 h to develop the anticoagulant effect. For a quick remedy, a heparin is prescribed concomitantly to cure patients with deep vein thrombosis and pulmonary embolism and to prevent emboli in patients with atrial fibrillation and mechanical prosthetic heart valves. The structure of a heparin molecule is shown in Fig. 7.3.

Acenocoumarol, phenprocoumon, atromentin, and phenindione are also in use. A synthetic sugar like fondaparinux, which contains five sugars (pentasaccharide) in heparin, binds to antithrombin; its structure is given in Fig. 7.4.

In bivalent drugs hirudin, lepirudin, and bivalirudin are direct thrombin inhibitors where the anticoagulant directly inhibits thrombin. Similarly, argatroban and dabigatran are monovalent drugs. The antithrombin protein is used as a protein therapeutic that is purified

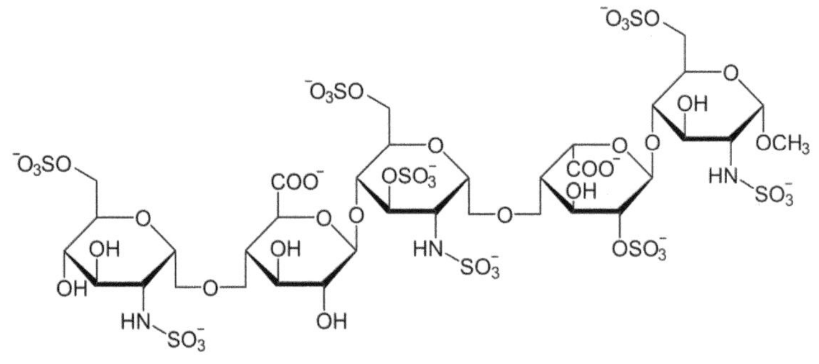

Figure 7.4 Structure of a fondaparinux molecule.

from human plasma or produced recombinantly (atryn, produced in milk of genetically modified goats). Also batroxobin is a toxin from snake venom that clots platelet-rich plasma without affecting platelet functions (lyses fibrinogen). Laboratory instruments, blood transfusion bags, and medical and surgical equipment get clogged up and become nonoperational if blood is allowed to clot. Test tubes which are used for laboratory blood tests have chemicals added to stop blood clotting. Apart from heparin, most of these chemicals work by binding calcium ions, preventing the coagulation proteins from using them. Ethylenediamine tetra-acetic acid (EDTA) irreversibly chelates (binds) Ca^{2+} to prevent blood from clotting. Citrate in liquid form is used for coagulation tests, as well as in blood transfusion bags. It also weakly binds the Ca^{2+}. So a stoichiometric proportion of this anticoagulant to blood is crucial because of dilution, and binding could be reversed with Ca^{2+}. Sodium citrate or acid-citrate-dextrose is used. Oxalate as an anticoagulant is used in fluoride oxalate tubes to determine the levels of glucose and lactate. Therefore, the friccohesity experiment could be performed by using suitable amounts of citric acid for anticoagulating activities.

Since the viscosity is directly proportional to the size of the integrated molecules, which act as a single body and create their own environment in a quantized state, the medium interactions are minimized so that no medium molecules could chemically or covalently bound or attach with the solute or the macromolecules used for molecular weight determinations. Such attachments induce chemical activity where the liquid–solid interface (LSI) senses the fluid flow and reads it.

7.2 Friccohesity Model for Determining Polymer Molecule Weight

The molecules are sensitive toward their own environment, and so the chemical environment must be reduced for determining the molecular weights, as is noted in Eq. 7.1.

$$[\eta] = k.M^{a} \tag{7.1}$$

The log of Eq. 7.1 is taken to develop Eq. 7.2, where $[\eta]$ is the intrinsic viscosity for a polymer solution, M is the molecular weight, and a is a constant for a polymer–solvent system. Thus it indicates that if the interaction between polymer and solvent is weaker then a is lower in values. Such theory is applicable in case drug-solvent interactions, salt-water interactions, and others. The log form of Eq. 2.4 is noted as under:

$$\log[\eta] = \log k + a \log M \tag{7.2}$$

The constant a is minimized to get a direct LSI sensing without any additional term. The survismeter produces accurate or M real values. The Newtonian medium assists in determining correct molecular weight, as no additional and localized chemical activities are generated in the sample during the measurements. Thus for molecular weight determination dv/dy (velocity gradient, s^{-1}) affects the result. dv/dy is a major rider for weight determination where the laminar layer could be equated to the surface with definite CFs. Thereby the friccohesity could also be taken into account for calculation of the molecular weight. In practical situations dy/dy decides the correctness of the weight determination and also the values of k and a constants. k is directly related with dv/dy, and a is related to the chemical activities in the medium. Thus new equations, Eqs. 7.3–7.5, could be formulated as under:

$$[\eta] \pm [\sigma] = k.[M^{a}. \pm \Delta\eta] \tag{7.3}$$

$$[\eta] \pm [\sigma] = k.[\pm\Delta\eta].M^{a} \tag{7.4}$$

$$[\eta] \pm [\sigma] = k.[\pm\Delta\eta].M^{a\pm\eta} \tag{7.5}$$

7.3 Experiment for Drug Loading by a Dendrimer

Determine the loading of a drug by a dendrimer in dissolving the dendrimer like trimesoyl 1, 3, 5 tridimethyl malonate (TTDMM) or trimesoyl 1, 3, 5-tridiethyl malonate (TTDEM) or sucrose monopalmitate (food nanoemulsion forming agent) in aqueous dimethyl sulfoxide (DMSO).

Theory: 0.001 to 0.020 drug (silibinin) solutions at an interval of 0.002 mol kg^{-1} dendrimer are prepared, and their intrinsic viscosity is determined using the survismeter. The comparative data are fitted in Eq. 7.2. Log M of aqueous dendrimers for dendrimers is fitted to calculate logs k and a. The values of these constants are used for M, which holds the drug. Such simulations determine the amounts of the drug encapsulated by the dendrimer. The medium allows molecules to flow, but the drug is encapsulated within the dendrimer and its fluidity changes, which quantitatively determines the amount of drug encapsulated therein.

Question: Determine iodine absorption by starch or TTDMM or TTDEM, similar to drug loading with a dendrimer in dissolving the dendrimers in aqueous DMSO.

In light of such observations monodisperse molecules give better results so dendrimers like TTDMM could act as standard. Also dv/dy with respect to CF is important and hence the molecular motions, which decide the impact of them on dv/dy, could be affected, which could be determined with dynamic light scattering (DLS). This view could be made with a coating of light active material to predict the states of the molecules. Thereby viscosity seems to be the correct method to get enough time to get optimized in lieu of their functional groups and geometrical dimensions because no other forces are applied except streamline flow of the molecules in dispersion. The survismeter brings the haphazardly moving molecules within the radii of 0.5 mm in parts so that there is a proper sizewise distribution of molecules. For example, initially the sample is transferred from a sample holder like a flask to the reservoir bulb (RB) of the survismeter, which is equilibrated for 30 to 45 min. depending on the nature of the molecules. Later the molecules are scanned out via a zero-energy band of 4 mm inner diameter (id) through a 25 mm compartment (controlled pressure unit [CPU]). This compartment acts as a global center and supports the local activities and bifurcates the sample for further scanning or distribution of the molecules for correct molecular weight determination. From the CPU to the functional bulb (FB) the scanning is made via <1 mm id. This gives one more opportunity to the molecules to get dispersed further for attaining a monodisperse state of the molecules. The distribution is made with respect to the FBs. Thirdly further scanning is made in buffer bulbs (BBs). These are the opportunities to the molecules to get them on a scale of monodisperse despite the riders of kinetic energy as their kinetic energy should be minimized when they are used for measurements with the survismeter solid surface so that the LSI becomes significant.

Chapter 8

Cross-Partitioning Law of the Survismeter

When backflow occurs the heavier molecules come first and hence the competitively active molecules remain operative on the flow. This further scans out the molecules for a better result of the molecular weight determination as no effective chemical changes occur due to a distance as per the Lennard-Jones potential (LJP). Such law is noted as the cross-partitioning law of the survismeter. It is also called virtual resistance for molecular partition.

8.1 Basics of Friccohesity Science

The fluid flow within the rigid Schrödinger boundary condition is monitored by several factors, which in a most synchronized manner control the flow in variable order, noted as under:

$$\frac{dV}{dt} = \frac{\pi r^4 p}{8\eta l} \tag{8.1}$$

Here, V is the liquid or sample volume permitted through a length l of the capillary, r is the inner diameter (id) of the capillary, p is the head-on pressure, t is the time, and dt is the time difference for the

Survismeter: Fundamentals, Concepts, and Applications
Man Singh and Sunita Singh

ISBN 978-981-4774-70-3 (Hardcover), 978-0-429-02761-1 (eBook)
www.panstanford.com

volume difference dV.

$$A = \frac{\pi r^4}{8l} \tag{8.2}$$

In this equation, π is a geometrical constant, r is the radius of the survismeter capillary, and l is the functional flow rate controlling length. On putting the A value from Eq. 8.2 in Eq. 8.1, Eq. 8.3 is obtained.

$$\frac{dV}{dt} = A\frac{p}{\eta} \tag{8.3}$$

The value of the head-on pressure p depends on the density of the liquid and gravitational forces g c/cm^2, as under:

$$p = h\rho g \tag{8.4}$$

Put the value of p from Eq. 8.4 in Eq. 8.3, as under:

$$\frac{dV}{dt} = A\frac{h\rho g}{\eta} \tag{8.5}$$

For determining the value of t from the equation, the equation is rearranged as under:

$$dV\eta = Ah\rho g dt \quad \text{or} \quad t = \frac{\eta}{A\rho g}\int_{h_1}^{h_2}\frac{dV}{h} \tag{8.6}$$

The final equation is obtained as under:

$$\eta = B\rho t \tag{8.7}$$

8.2 Solubility Determination with a Survismeter

Experiment: Determine the solubility of the given sample in a given solvent using the survismeter.

Procedure: The solute is added in a cumulative manner, and the viscosity keeps increasing in a linear manner, which indicates the solubilization of the solute in the given solvent at a given experimental temperature and pressure. Initially, the calibration curve is determined for the value of regression analysis of the viscosity versus concentration data, as under:

$$\eta = \left(\frac{t}{t_0} \times \frac{\rho}{\rho_0}\right)\eta_0 \tag{8.8}$$

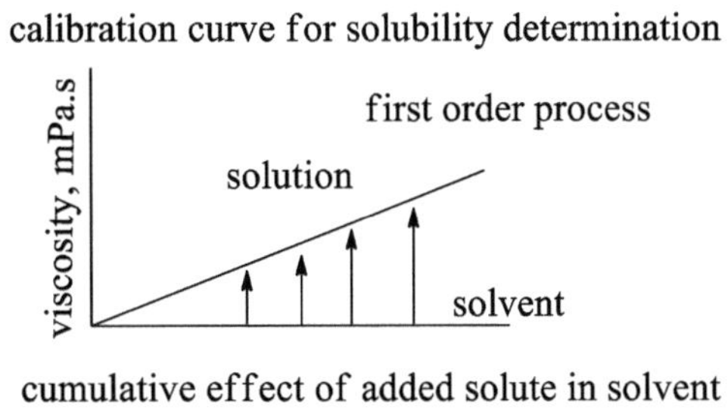

Figure 8.1 Nonideal liquid systems depicted by an increase in viscosity on account of interactions.

Here, t and t_0 are the viscous flow times, ρ and ρ_0 are the densities, and η and η_0 are the viscosities of the solutions and the solvent, respectively. By determining the viscosities as a function of the compositions c the following standard equation could be fitted.

$$\eta = \eta^0 + A_c \text{ (calibration standard equation)} \quad (8.9)$$

η^0 is the extrapolation constant and A is the slope constant for a given composition of the sample solution. These data are used for determining the solubility of an unknown sample by determining the viscosity of the unknown sample.

$$\eta = \eta^0 + A_c \quad (8.10)$$

The calibration curve depicted in Fig. 8.1 can be used to determine the surface tension of urine samples of persons of ages from 5 to 70 years. Such experiments can act as health indicators as the results enable to identify the state of uric acid or sugar content in the body.

8.3 Biophysics Experiments for Testing Blood Sugar

A calibration curve developed with a survismeter for checking the blood sugar of a normal person could act as a standard for determining the sugar of a sugar patient.

***Procedure*:** 0.5 mL of blood is dispersed in a buffer as a dispersing medium, and surface tension and viscosity both are determined for a calibration curve, which is extended for the unknown sample. Similarly, the survismeter determines the number of carbon atoms in a surfactant alkyl chain. The surfactant determines the hydrophilic head group of the surfactant, like quaternary nitrogen (QN), phosphate, nonionic, and cationic through friccohesity. The friccohesity accurately determines the cohesive force (CF) and the frictional force (FF) or the degree of solubilization, the degree of nanoemulsification, and other structural changes. The physicochemical properties (PCPs) reflect the behavior of the molecules when they are subjected to variable experimental and industrial uses. Their data reflects the following:

- Compatibility of molecular as well as capillary geometries
- Nonenthalpic expression of molecules within entropic scale
- Ionic-hydrophilic-hydrophobic (IHH) expressions
- Friccohesity and tentropy of mixtures
- Chemistry of physicochemical sciences for industrial nanoemulsion
- Introduction and foundation of short-term course
- Origin of the capillary concept vis-à-vis the molecular concept within a 1D scale
- Robert Boyle concept of partial pressure
- Leonardo da Vinci capillary theory and Poissellis viscous liquid flow
- Linear to aligned molecular orientation in an experiment
- Activation energy in a nontransition state
- Intramolecular multiple forces theory (IMMFT) and tentropy-driven molecular activities
- Nanoemulsion and nanoparticle capping activities
- Adsorption and desorption activities
- Determination of coalescence and coagulation
- Differentiation of emulsion and solution preparations using the survismeter

Chapter 9

Intramolecular Multiple Forces Theory

Macromolecules, like proteins, with active peptides bonds that are embedded in internal structures (alpha and beta sheets) show interacting activities without breaking real bonds. This type of process is known as an entropic process because many interacting electrostatic sites remain operative and work. Such working mechanism works on the basis of the intramolecular multiple forces theory (IMMFT). During an IMMFT implementation internal arrays of variable entropic orders are generated on the basis of the Boltzmann distribution patterns. These internal entropic orders are called tentropy. With such effective activities friccohesity is an operator that develops an interacting interface of protein-like molecules with solvent, leading to their diffusion or dissolving in the solvent. The cohesive forces (CFs) of proteins on interaction with solvents are transformed into viscous forces or frictional forces (FFs). So friccohesity is a creative physicochemical property of macromolecules like proteins or webdrimers and dendrimers that depicts their unfolding activities.

Survismeter: Fundamentals, Concepts, and Applications
Man Singh and Sunita Singh

ISBN 978-981-4774-70-3 (Hardcover), 978-0-429-02761-1 (eBook)
www.panstanford.com

9.1 Unique Theory of New Science: Friccohesity

CFs are geometrical optimization of anti-Gaussian factors, and this approach leads to the development of the concept of friccohesity. For example, the friction on the way induces molecular activities whereby the molecules, due to the friction, induce chemical activities causing molecular motions like rotational, translational, vibrational, electronics, and torsional forces. Had there been no friction or had the medium not had the chemical forces there would have not been an interconnection for equilibrium distribution of the force factors of the medium. Thereby, varieties of molecular motions would have not caused a networking. Since molecular motions help the molecular active sites to respond as per their numbers and potential energy, the CFs involve Lennard-Jones potential (LJP), ionic, molecular forces, and geometrical activity–driven forces and the FFs involve molecular activities. Thereby, one can say that friction is a two-path process like hysteresis, redox, and charging-discharging chemical activities. Thus, the spatial and geometrical molecular activities are most wanted in the case of chemical kinetics for developing transitional complexes of the activated species. For example, heating the water from ice to vapor state and further cooling the vapor back to ice could be noted as an ideal system to follow the correct heating and cooling curves because there is no chemical transition on the way (Fig. 9.1).

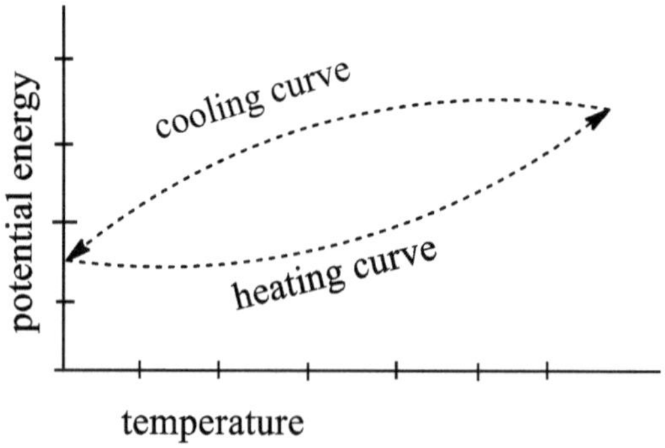

Figure 9.1 Thermodynamics of temperature driven molecular activities in terms of their potential energy.

Molecular forces due to friction accommodate and in search of accommodation vis-à-vis solvent resistivity, they undergo charge transfer, hydrogen bonding (HB) disruption, and philic–phobic activities or micelle formation. The FF becomes the most wanted effective factor for aggregating or coagulating to a suitable geometry. These are the reasons that surfactants and salts show different behaviors—because of their approach vis-à-vis solvent and hydrogen-bonded networking with a definite chemical potential of the medium. Thus the approach and magnitude of the chemical potential of the solute or the adsorbate matter a lot. Such visible and working science establishes the significance of the friccohesity theory. The evaporation of the solvent also causes friccohesity because the molecules that gain energy do show separate behaviors whereas the molecules that are yet to gain energy are in a different energy state. Thus both energy states induce a separate response. The FFs are caused due to the medium and also the tendency of other components or adsorbate or additive in the solvent. Two sets of force coordinate work between the solute/adsorbate and solvent, which optimize aqueous solutions like immiscible solvent systems. Immiscible solvents can be further mutualized by the surfactants. Once the CFs are weakened then the individual molecules move poles apart and in the process of moving poles apart, the molecules attract the dissimilar molecules added for making the liquid mixture.

Dissolving chemical species in specific solvents disrupt the HB of the solvent for preparing a homogenous liquid mixture for an equal solute distribution, verified by measuring the friccohesity. A disruption of hydrogen bonds of solvents like water reduces both the cohesivity and the CFs. The pure water as a solvent possesses 100% hydrogen-bonded structure, with the CFs to their fullest strength. On dissolving the citric acids, metal chlorides, and ionic liquids (ILs) the hydrogen bonds are disrupted, where the CFs are partially transformed into molecular forces among the dissimilar molecules, which are rightly defined as intermolecular forces (IMFs). The percentage of transformation of CFs into IMFs or vice versa leads to the existence of friccohesity as an operator or interface locater of the solvent and the solute. Thereby the concept of friccohesity is most specific for developing a homogenous mixture that critically defines an actual state of the solvent–solvent and solute–solvent interactions with the CFs and IMFs, respectively. So solvent–solvent and solute–solvent interactions have been in focus for explaining the friccohesity processes of utmost importance.

9.2 Friccohesity for Understanding Colloids

Friccohesity is helpful in understanding the colloids that are translucent because of the Tyndall effect, which is the scattering of light by the particles in colloids. Some of the colloids may be opaque or have a slight color. The colloidal suspensions develop interfaces, and colloid science, which was introduced in 1861 by Scottish scientist Thomas Graham, has emerged as a thrust area of research. The dispersed-phase particles have a diameter of 1 nm $\approx$ 1000 nm. Such particles are visible under an optical microscope, although at the smaller size range ($r < 250$ nm), an ultramicroscope or an electron is needed. For example, bentonite is an absorbent that is aluminum phyllosilicate clay consisting mostly of montmorillonite. It was named by Wilbur C. Knight in 1898 after Cretaceous Benton Shale near Rock River, Wyoming. The bentonites are named after the respective dominant elements, such as potassium (K), sodium (Na), calcium (Ca), and aluminum (Al). Experts debate a number of nomenclatorial problems with the classification of bentonite clays. Bentonite forms from weathering of volcanic ash, most often with water. The term "bentonite" and similar clay, called tonstein, describe clay beds of uncertain origin. For industrial purposes, two main classes of bentonite exist: sodium and calcium bentonite. In stratigraphy and tephrochronology, completely devitrified (weathered volcanic glass) ash-fall beds depict K-bentonites when the dominant clay species is illite. Apart from montmorillonite and illite another common clay species dominant is kaolinite. Kaolinite-dominated clays are commonly referred to as tonsteins and are typically associated with coal.

Thus, the FF is most useful in bringing the adsorbate into the nearest contact of the solvent or other additives, like the catalyst, which also gives time to reorient the surfactants, drug, and adsorbates.

It is seen that a flight of molecules in secondary structure as a soft medium develops valuable physicochemical properties that reflect internal structural changes. The secondary structure is made up out of suitable HB involving energetics within 10–40 KJ/mole with a definite IMMFT, which is applicable in case of the giant molecules like proteins and functional polymers. These molecules have multiple interacting sites within the structure of a single molecules. The secondary structures are the fundamentals of the solvent and solute sciences where the solutes do have HB developing sites, which coordinate with counterparts of the solvents. The HB coordinating sites have 1:1 coordination patched up with electrostatic energy, geometry, dipoles moment, and others. Such noncovalent changes are studied with the survismeter in terms of the friccohesity, which is very functional in developing the nanoemulsion of the hydrophobic and hydrophilic substances. The nanoemulsions are effective in dissolving several materials. Some important facts related to nanoemulsions are as follows:

- Nanoemulsions dissolve e-wastes and polymer wastes.
- Reacting and interacting nanoemulsions are known as kinetic and dynamic nanoemulsions.
- Nanoemulsions dissolve piercing nanoemulsions and interfacial nanoemulsions.
- Percolating nanoemulsions are known for decoating and removing ability.
- Nanoemulsions help in removing and peeling off paint.
- Similar to adsorption of CH_3COOH by activated charcoal the series of CH_3RCOOH (R = CH_3CH_2-, $CH_3CH_2CH_2$-, $CH_3CH_2CH_2CH_2$-) could be studied.
- Also it could be tried for phenols, amines, and alcohols.
- Experiments similar to the titration method can be conducted with the survismeter.
- Total available surface area = 100, θ covered area of adsorbent $= 1 - \theta$
- Also a surface excess concentration could be studied.

- A graph could be plotted between the surface energy and the surface area a of the surfactant.
- The interacting constant of NaCl in aqueous solutions could be determined by measuring the friccohesity using the survismeter.
- The glucose binding constant in aqueous solutions could be determined by measuring the friccohesity using the survismeter.

Basic science of drying food involves enhancing binding forces. Water is the most effective agent for inducing friccohesity; it enters the embedded structures and induces intramolecular multiple forces.

Chapter 10

Friccohesity Determination with the Survismeter

Friccohesity involves a multistage relaxation process. The relaxation stages and steps ensure regaining and optimizing of the natural state of molecules either in the cohesive cages or in the frictional cages. The cohesive cages are divided into two categories, the positively charged and the negatively charged cages.

10.1 Normal Cohesive Cages

The normal cohesive cages (NCCs) belong to the linkages of similar molecules, like water and water. Such state belongs to the pure solvent when the solvent molecules as per their electrostatic or London dispersive forces (LDFs) gain a structural state with definite physicochemical properties. Cohesive cages are further extended indirectly, such as in the formation of ionic hydration, capping of nanoparticles (NPs), and micellar systems. In these systems the solutes, like salts, disrupt the solvent hydrogen bondings (HBs) and realign the charges of the solvents, like water, having negatively charged O^{2-} and positively charged H^{+}. These charges freshly

Survismeter: Fundamentals, Concepts, and Applications
Man Singh and Sunita Singh

ISBN 978-981-4774-70-3 (Hardcover), 978-0-429-02761-1 (eBook)
www.panstanford.com

realign around the cations and anions on dissociation of the salt molecules. Thus in the secondary structure the formation of enthalpy is effectively involved or the structural thermodynamics becomes the most deciding factor. Thus the secondary structural systems could be termed as HB disruption and realignment (HDR)-driven electrostatic systems. Thus considering both the NCC and HDR systems the survismeter measurements for friccohesity induce relaxations that could be named as under:

- Activation of molecules in NCC and HDR
- Initial relaxation after filling up the sample
- On evacuating air pressure from air liquid interface
- Filling sample in controlled pressure unit (CPU) of the survismeter
- Extension 01: Extending filling to functional bulbs
- Extension 02: Extending from functional bulb to buffer bulb (BB)
- User relaxation on withdrawing piston arrangement
- Disintegration with uniform rate
- Soft phase equilibrium and aligning relaxation
- Pressure relaxation for streamlining

In light of understanding the above points the natural structure of the liquid mixtures could uniquely respond to the external forces. The measurements made with the survismeter do not involve application of any external forces, like optical, mechanical, and electromagnetic, because such forces could induce structural deformation. So such forces are avoided in the experimentation until and unless they are needed. These forces induce coalescence, coagulation, aggregation by causing stretching, vibration, and rotational and translation thermodynamic motions with respect to the friccohesity of the liquid mixtures. Therefore, the survismeter becomes a most relevant technique to determine such changes occurring in the liquid mixtures. These effects disturb the free radical mechanism and the drug binding and release process. Therefore, the survismeter methodology is adopted for the study of the molecular interaction involved in formulations.

10.2 Supporting Reasons for Measuring Friccohesity

In surface forces or energy or tension the liquid has inward pull to attain an optimized state. Such tendency develops the pendant drop number (pdn) when the liquid is permitted certain surface area of a cross section of capillary. Interfacing of the area of the cross section (πr^2) and cohesive forces (CFs) of the liquids facilitates determination of the surface tension using the survismeter. Friccohesity is an electronic theory that indicates electronic interactions. It deals with all interacting sites or moieties of individual molecules within the boundary of a molecule. Thereby, an essential component of friccohesity is the Schrödinger boundary condition and Boltzmann distribution theory. The molecules are bound together with residual forces with a definite electron cloud. Friccohesity explains electronic cloud interactions in liquid mixtures.

10.3 Salient Features of Survismeter Science

- Origin of physiochemical sciences studied with the survismeter in $\Delta G < 0$ mode
- Chemistry of electronegativity, delocalization, functional groups, and lone-pair electrons inducing physicochemical activities responsible for the nanoemulsion
- Philic-phobics for interfacial dynamics of biphasics
- Solvodynamics to statodynamics of mutually homogenizing thermodynamically
- Friccohesity of nanoemulsion forming components
- Structure breaking and making mechanism in $\Delta G < 0$ mode
- Nanoemulsions and interfacial thermodynamics
- Fortified food for balanced intake and sustainable requirement

Fundamentally nanoemulsions are based on the transformation of the potential energy of oil or water into developing their ninilets or nanolets (let = local equilibrium thermodynamics) together. Such sciences of the oil or water develop spontaneity for the placement of oil or water in core or peripheral positions. For nanoemulsions, Gibbs energy (ΔG, J $mol^{-1}K^{-1}$) is positive and attributed to either the oil or the water phases, which are disrupted and forced to be involved in mutual solubilizing arrangements as per their amounts present in the nanoemulsion. The potential or binding energy responsible for holding the oil or the water phase together now is transformed into kinetic energy, where the nanolets induce Robert Brownian motion or could be noted as friccohesic motions. $\Delta G > 0$ is evolved because the oil and water phases are to be dewetted into nanoemulsions due to the formation of nanolets. By mechanical shaking, stirring, sonication, and other methods, possibilities are explored to disrupt their bulk phases by dechannelizing their binding forces, like van der Waals and LDFs, mutually so that the emulsions remain stable. The molecular forces develop physicochemical properties which infer the resultant state of the liquid mixture noted as follows:

- Nanoemulsions and their applications for health and in routine working
- Understanding the individual binding energies in bulk phases
- Factors for binding energies and mutual solubilization
- Distinguishing nanoemulsions from solutions
- Distinguishing the viscosity of a nanoemulsion from that of a simple emulsion
- Concept of emulsion, microemulsion, and nanoemulsion
- Constituents of best emulsion formulation
- Concept of their stability via attaining adequate binding forces
- Structural interlocking for least energy attainment
- Additives and their mechanism to solubilize or be dispersive
- Basis of their applications of nanoemulsions
- Heat holding capacity and ultraviolet (UV) light absorbing ability
- Water saving approach and food value adding science of nanoemulsion
- Particle sizes and their functions with chemical additives

Chapter 11

Food Molecules and Calorie Value

One of the deciding steps in a nanoemulsion is to develop nanolets that are interfaced with surfactants and the medium. The nanoemulsion incorporates the solution chemistry or the nanosolution chemistry (nanosolchem), involving nanothermodynamics, nanofriccohesics, and biopharmaceutical research, as the advanced sides of nanoscience and nanotechnology. Thereby, interconnecting activities develop new nanofriccohesics incorporating surfactants, DNA templates, globular proteins, hormones, and similar others, which constitutionally involve friccohesity in molecular force development or distribution within electronically bound molecular structures (EBMSs), which emerge from the atom-to-atom contact via delocalization or electrostatic or proton transfer, like zwitterions, induced by pH. The EBMSs control transfer of the aqueous solutions on a scale of change in chemical potential and cohesive surface energy whose uncertainties could negatively affect the precision of operation either to have safer dealing with hazardous reagents as a by-product right down the proverbial drain or to have fruitful end products. The concepts help understand or implement the understanding and mechanism both in day-to-day working and for sustainable life as the need of living beings. Thermodynamics and

Survismeter: Fundamentals, Concepts, and Applications
Man Singh and Sunita Singh

ISBN 978-981-4774-70-3 (Hardcover), 978-0-429-02761-1 (eBook)
www.panstanford.com

friccohesity are action-oriented sciences and if followed could help develop a good life.

11.1 Nanoemulsion for Anticorrosion, Antifat Deposition, and Fat Hydrolysis

The molecular interaction study using the survismeter determines the friccohesity data, which reveals the state of a liquid mixture. The boundary conditions are the most relevant theory for the study of nanoemulsions or microemulsions for characterization via fluid dynamics using the survismeter. The molecular forces within the molecular structures undergo a change within the close distance so that the van der Waals forces, the Lennard-Jones potential (LJP), and the London dispersive forces (LDFs) could operate. Such forces for study need certain conditions, like allowing them to flow within a capillary of a fixed inner diameter (id) having stationary and rigid boundaries. The nanoemulsions develop Brownian motions, including other molecular motions, which are much affected by boundary restrictions. For nanoemulsions, the survismeter, with a uniform id and bends at 45°, normalizes the molecular motions to reproduce data. The survismeter quantitatively tracks their interfacial tension (IFT), surface excess concentration, surface energy/tension, viscosity, particle sizes, friccohesity, tentropy, and intramolecular multiple forces theory (IMMFT). The id of the survismeter remains within the condensed matter range, where the molecules are in close proximity and are able to exchange electron clouds, maybe in the form of exchanging charge or energy, on an interaction scale from 10 to 40 KJ mol^{-1}. Since the rigid capillary boundaries trim the flow pattern of molecules in solvents, interaction of molecules with solvents develops unique patterns of dynamic viscosity. The dynamic viscosity depicts a fluid that shows effective dynamics within the capillary. So it is not a static but a dynamics process in nature because on viscous flow the molecules could realign. The survismeter has provisions to trim and track molecules within the rigid capillary boundary having $\pi r^2 l$ (π is a constant, r is the id of the capillary, and l is the length of the capillary) control. It also has the nonboundary condition in the case of pendant drop formation, where molecules override each other in the form of layers to develop pdn adhering to the lower tip of the surface tension measuring limb. However, similar to size exclusive chromatography (SEC) or gel permeation chromatography (GPC) a differential or velocity gradient could be applicable in the survismeter capillary despite a narrow id. The pdn theory is based on the following relation:

$$2\pi r\gamma = mg \tag{11.1}$$

Here, π is a constant; r is the survismeter constant as the id of the capillary, which is fixed for specific equipment; and γ belongs to the liquid sample. The 2 depicts two sides of surface faces under stretch as a stretched elastic membrane due to hydrogen bonding (HB) in continuity. However, in the case of Hg the stronger covalent forces are responsible for higher surface tension. Equation 11.1 explains that the higher the γ, the larger is the weight of a pendant drop. The r as the id faces hydrostatic pressures so its relation with pdn is most critical and strategic in nature. It has direct relation with the surface tension of the liquid, and it is normalized with surface tension. Though in the case the dynamic viscosity measurements, mass transfer take places, which is in continuity applying frictional forces (FFs), in the case of surface tension measurements the FFs are ceased off and adhesive forces take over at the tip of the capillary. For viscosity and surface tension measurements, activation energy is involved. An adequate amount of energy could be generated out of molecular motions being triggered in the mixtures during measurements. It has specific science at the outset of the liquid taking position to flow that needs reorientational changes acquiring energy noted as activation energy.

11.2 In Situ pH Impact of Phenols

Experiment: Determine in situ pH impact of phenols (phenol, 2-hydroxyphenol, and 1, 3, 5 trihydroxyphenol) (Fig. 11.1) in their aqueous solutions on measuring friccohesity, surface tension, and viscosity.

Procedure: The phenols as weak acids liberate protons. The protons could affect cohesive forces (CFs) of their aqueous solutions.

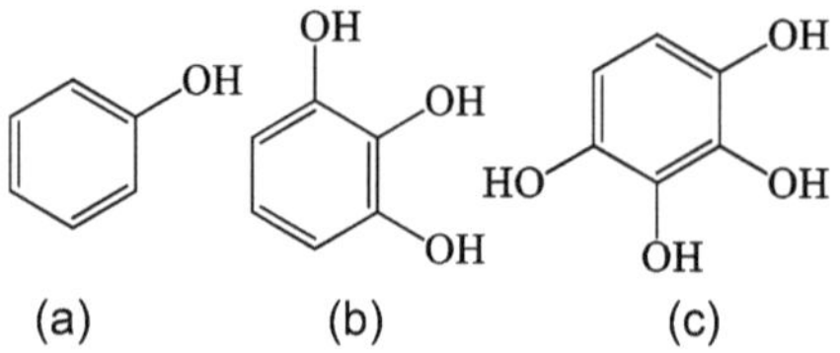

Figure 11.1 Molecular structures of (a) phenol, (b) 2-hydroxyphenol, and (c) 1, 3, 5 trihydroxyphenol.

The friccohesity, surface tension, and viscosity for 1.0 to 4.5 mol kg^{-1} aqueous phenol solutions are measured at the desired temperature. Since the molalities of the phenol aqueous solutions remain the same, variations in their data infer quantitative effect on pH of their aqueous solutions.

11.3 Fat Dissolution by a Eutectic Mixture of Choline and Urea

Both choline chloride and urea (Fig. 11.2), in 1:2 stoichiometric ratios, develop a most significant deep eutectic mixture. The mixture has a melting point of 12°C, less than the melting point of choline, 302°C, and urea, 133°C. The friccohesity of such mixtures could be most useful.

11.4 Activation Energy

Activation energy is evolved out of the flow of ions or molecules in Brownian or molecular motions. Such motions could induce

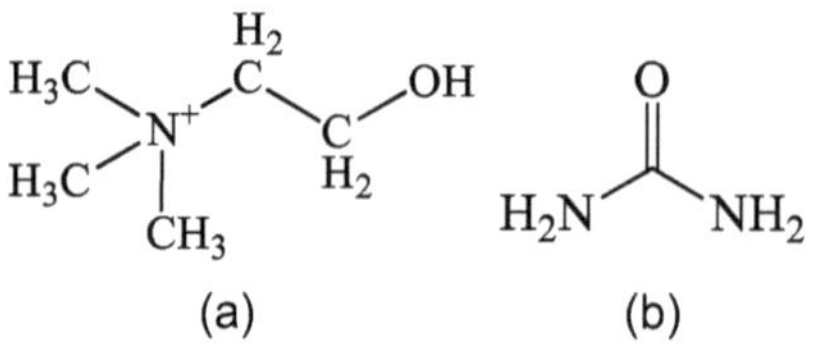

Figure 11.2 Molecular structures of (a) choline and (b) urea.

arrays of specific types, like isotropic and anisotropic. The molecular motions push the ionic hydrated arrangements to one direction, which causes potential deference and develops a current to charge a cell phone. The osmotic pressure and charger are fitted because the process is dynamic with kinetic energy. These mechanisms flip-flop the ions to move further ahead with specific velocity and charge alignments of phenols.

11.5 Relationship of Kinetic Energy with $\pi r^2 l$ Condition

Viscosity and surface tension, as the most preferred physico-chemical properties of liquids, are measured together. Viscosity and surface tension authentically identify Brownian motions and coalescence processes with specific particle sizes and surface areas, which affects the overall physicochemical profile of liquid mixtures. It becomes most useful when the molecules develop movement and could lead such a scenario to have a "mean free path." This behavior of molecules could act as a sensor in the presence of several solvents and chemical additives. So glycine, with amino ($-NH_2$) and carboxylic (-COOH) functional groups, along with methylene ($-CH_2-$) groups, develops variable friccohesity with different solvents, pH, and temperatures and with surfactants, emulsifiers, salts, and carbohydrates. The kinetic energy of molecules depicts a response of the solute with solvents because when they are mixed together they share their molecular forces, generated from their molecular structures. Figure 11.3 illustrates the molecular distance and the rigid capillary id range. The liquid samples have flow within the rigid boundaries so the pressure generation plays a useful role in the measurement of data using the survismeter.

So pressurelets (virtual valves for surface tension, viscosity, and pressure limb) are made inside controlled pressure unit (CPU) of the survismeter for diverting pressure and liquid in the same direction. The survismeter engineers pressure with respect to dimensional designing for safer and authentic bypassing of the sample as per the measurement sequence. The emulsion refers to the photosensitive

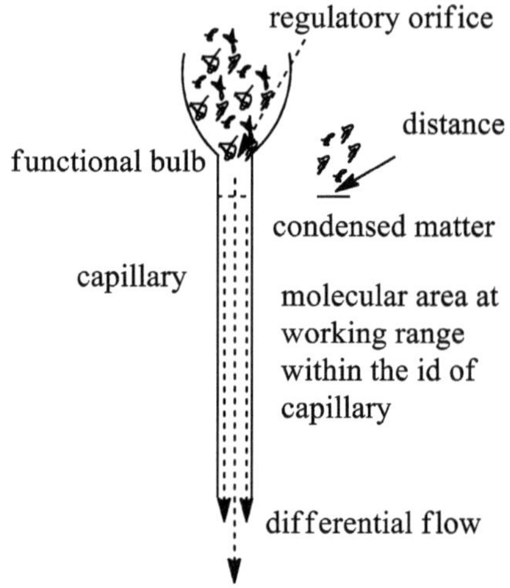

Figure 11.3 Dynamics of viscous fluid flow within the rigid and uniform capillary of survismeter.

side of photographic film, noted as photographic emulsion. Such emulsion contains silver halide (AgX, X = halide anion) colloidal particles dispersed in a gelatin matrix. Friccohesity measurements of such liquid seem to be the most kinetic, dynamic, and informatory. The survismeter could be useful for nuclear emulsions, which are similar to photographic emulsions but are used in particle physics to detect high energy. The survismeter could have several modes or approaches to deal with and to explain the data, which is noted as the EBMS model. The EBMS induces and causes n number of orders of molecules on flow in the manner noted in Fig. 11.4.

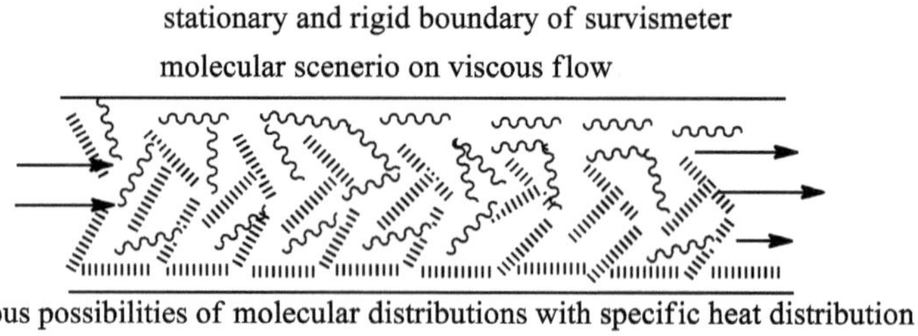

Figure 11.4 Internal molecular motions, atomic oscillations to optimize the viscous flow times as per nature of molecules of sample.

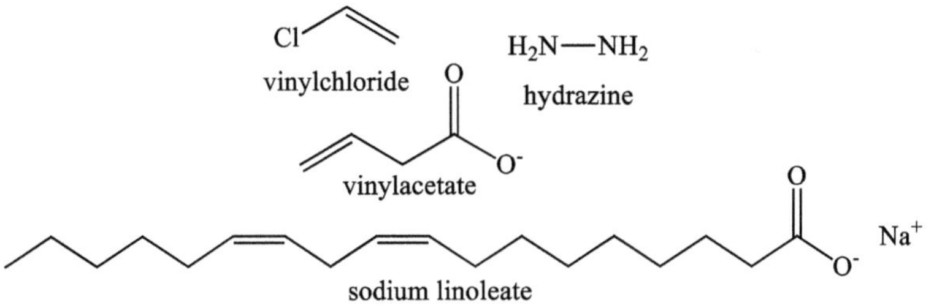

Figure 11.5 Molecular structures of constituents of liquid sample to investigate their mutual interactions.

For example, if the molecular structures given in Fig. 11.5 are mixed and allowed to flow through a narrow id capillary, several orientations could emerge on flow. The geometry and functional groups that are attached at specific positions also affect the geometrical expressions of the molecules.

Friccohesity (σ) for structure making or ionic solutions and dispersion is related with density (ρ), which is noted as under:

$$\sigma_{\text{aq-SDS}} > \sigma_{\text{aq-NaCl}} \quad \text{but} \quad \rho_{\text{aq-NaCl}} > \rho_{\text{aq-SDS}}$$

Figure 11.6 depicts a relationship between the density and the friccohesity.

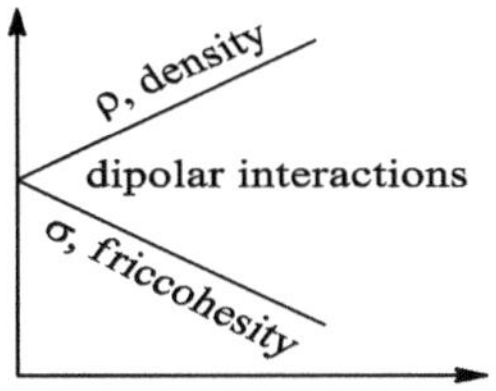

Figure 11.6 Relationship between density and friccohesity.

Chapter 12

Heat Carrying Emulsions: Lower Viscosity and Higher Surface Tension

A coolant as a fluid flows through or around a device to prevent the device from overheating. It transfers the heat that is produced by one device to another device. Such approach either uses or dissipates the heat that is produced in the process. An ideal coolant has a high thermal capacity, low viscosity, and a higher surface tension and is low cost, nontoxic, and chemically inert. It neither causes nor promotes corrosion of the cooling system. Some applications require the coolant to be an electrical insulator. The survismeter has become a most relevant device for measuring such parameters of coolants. The study of binding energy of ions using salts in a prescribed medium on integrating data infer an electrical ionic field.

- **Two-Way Hydration or Ionic Hydration**
  - Monovalent, like NaCl, and with a medium of prescribed polarity
  - Divalent, like $CaCl_2$, and with a medium of prescribed polarity
  - Trivalent $FeCl_3$ and with a medium of prescribed polarity
  - Study of ionic fields with friccohesity of aqueous salt systems

Survismeter: Fundamentals, Concepts, and Applications
Man Singh and Sunita Singh

ISBN 978-981-4774-70-3 (Hardcover), 978-0-429-02761-1 (eBook)
www.panstanford.com

- Structure making and breaking determination
- Distinction between micelles and coagulation study
- Quaternary valent $SnCl_4$ and with a medium of prescribed polarity
- Competitive hydration or zwitterionic hydration
- Quaternary nitrogen atom–derived properties
- Single-way hydration or surfactant-based hydration

- **Molecular Hydration Rigid Bond and Flexible Bonds**
  - Factors of hydration and hydrodynamic size or radii
  - Dipolar/electrostatic interactions with mediums of like polarity and ionic strength
  - Induced dipolar or intra-atomic electron shift–based interactions
  - Zwitterionic interactions with a medium of prescribed polarity
  - Globular protein–unfolding study with a medium of prescribed polarity
  - Dendrimer drug–loading study with a medium of prescribed polarity
  - Study of adsorption by adsorbents

- **Formulational Study**
  - Coalescence with a medium of prescribed polarity
  - Coagulation with a medium of prescribed polarity
  - Turbidity with a medium of prescribed polarity
  - Flocculants with a medium of prescribed polarity
  - Water-binding capacity of soil particles and water-binding capacity of gels or sols

- **Practical Applications**
  - Water-binding chemical compounds glycerol and NaCl used in sustaining bouquet or the twigged flower in a pot
  - Friccohesity-driven chocolate-making industry for maximum bioavailability
  - Colloidal nanolets in food items and pollutant removal through nanolets
  - Interfacial energy based on nanoemulsion application

12.1 Hydrophilic and Hydrophobic Domains

Currently new research areas, like food, beverage, and pharmaceutical, in utilizing edible nanoemulsions to encapsulate, protect, and deliver lipophilic functional components, like oil-soluble flavors, vitamins, preservatives, nutraceuticals, and drugs, are emerging. Several potential advantages of using nanoemulsions in place of ordinary emulsions greatly increase bioavailability of lipophilic substances. Nanoemulsions scatter light weakly and are incorporated into optically transparent products. The nanoemulsions effectively modulate the product texture and induce high stability against particle aggregation and gravitational separation. But there may be some risks associated with oral ingestion of nanoemulsions, such as their ability to change the biological fate of bioactive components within the gastrointestinal tract and the potential toxicity of some of the components used in their fabrication. This tutorial review provides an overview of the current status of nanoemulsion fabrication, properties, and applications, with special emphasis on systems suitable for utilization within the food industry. Currently, nanoemulsions are applied in areas of food, beverage, and pharmaceutical industries due to various potential advantages over conventional emulsions for wider applications.

12.2 Interfacial Thermodynamics and ΔG

Nanoemulsions greatly increase the bioavailability of highly lipophilic substances encapsulated within them, and so they are designed to have high stability against particle aggregation and gravitational separation. A nanoemulsion extends the shelf life of valuable products, and droplets in nanoemulsions are so small that they only scatter light waves weakly. These are the reasons that the droplets are incorporated into optically transparent products, like fortified soft drinks and waters, without causing any adverse effects on their clarity. A nanoemulsion could be highly viscous or gel-like when the droplet concentration is lower than that of the emulsion and could therefore be used to create products with novel textural attributes or reduced calories. Oil-in-water nanoemulsions

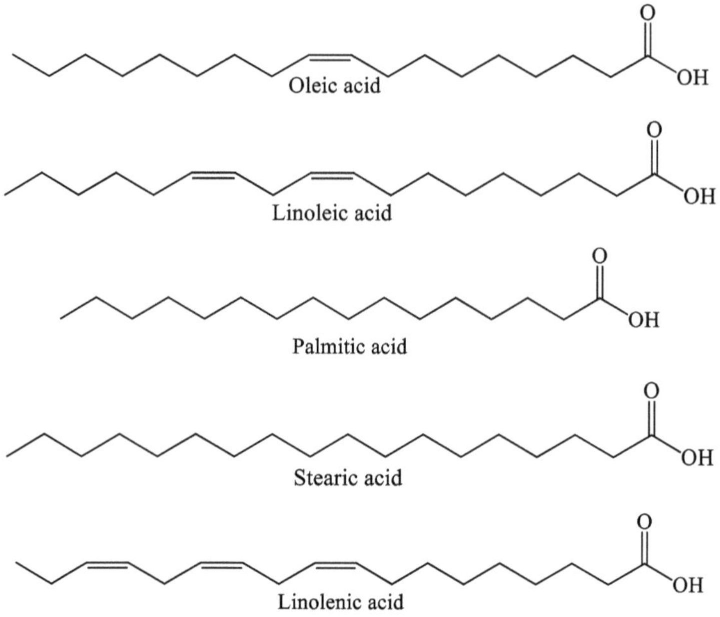

Figure 12.1 O-water nanoemulsions with increasing number of carbon atoms and increasing number of pi bonds in alkyl chain of oils.

are considered as the best minilet (mini localized equilibrium thermodynamics) mixture because of their greatest potential for application within commercial products. The oils listed in Fig. 12.1 could be used for such purposes with suitable surfactants like sodium dodecyl sulphate (SDS), DATB, cetyl-trimethyl-ammonium bromide (CTAB), and tweens, along with additives like polyethylene glycol (PEG), glycerol, ethanol butanol, guar gum, sucrose, and honey. Triglycerides with the following acids could be effective in developing fortified foods. It is seen that the oils have developed nanoemulsions with ethanol, glycerol, surfactants, and PEGs (Fig. 12.1).

12.3 Mechanistic Understanding of IFT of Salinity Effect

Mechanistic understanding of the effect of temperature and salinity on the water/toluene interfacial tension (IFT) explains mutual solubilities and the effects of salt and asphaltenes on IFT. Probably a closer scientific understanding of IFT and variations therein could be explained effectively in light of the friccohesity of the immiscible solvents. A few comments are gathered to further elaborate the explanations. So a new hypothesis is recommended for making new models of IFT for wider sets of immiscible solvents and solubilizing agents. The objective of integrated efforts and the current text is to present a mechanistic understanding, from an atomic level, of effects of temperature and salinity on IFT in the presence and absence of organic materials, like functionalized graphene and asphaltenes, at the oil–water (O/W) interfaces. So elaborated views of new interfaces signify the study and propose new openings. Asphaltene fractions stabilizing O/W emulsions could be a most pioneering work of mutual solubilization of the water-asphaltenes-organic phase. Such a model could develop additional cohesive forces (CFs) or intermolecular forces (IMFs), which are responsible for increasing or decreasing the IFT.

Therefore, in order to study the salinity effect, a specific number of water molecules could be replaced by Na^+ and Cl^- ions to achieve the salt concentration in an aqueous phase. The added Na^+ and Cl^- ions are of equal amounts to keep the simulated systems electronically neutral. It is a well-known fact that the Na^+ and Cl^- ions realign vis-à-vis water molecules on increasing CFs and weakening IMFs. So a critical link of the NaCl with asphaltenes at the water and toluene interface is needed. It should be mentioned that this newly developed thermostat is based on correctly reproducing the distribution of kinetic energy under a constant temperature. To validate the IFT trends observed in studies, several experiments for pendant drop measurements with the survismeter could be conducted on toluene solution/water binary systems. To probe the effects of temperature and salinity the solutions were made with toluene for measuring the IFT. For the IFT measurements, the survismeter is found to be the more authentic device than a tensiometer as the latter is unable to avoid vibrations, air fluctuations, temperature, and entropic shocks that vibrantly disturb pendant drops. Thereby, the survismeter authentically eradicates such effects that are involved in the study of IFT, especially at varying temperatures.

For taking note of hydrogen bonding (HB) in water, it requires a larger amount of energy to create a unit area of free water surface. Thereby, the number of hydrogen bonds among water molecules at different temperatures could be calculated, which could be normalized with respect to the number of water molecules. In this context how friccohesity acts as a deciding factor becomes significant because the CF in water is transformed into water-toluene. Both temperature and inorganic salts result in an increment in IFTs. This finding is similar to that observed for IFTs of water/organic solvents to understand the effect of salt ions. Hence there is a need to choose the working coordinates, like ionic field, of ions and ionic bonds with water and others. So there is need to throw light upon it and to actualize or normalize the increments in IFT, with salts realigning the water molecules with higher CFs and lower friccohesity data. Such sciences very specifically take note of the following:

- IFT needs a closer explanation and evolution of the science of the mutual solubilization of immiscible phases.
- What is the significance of a liquid–liquid interface (LLI) (water–toluene, water–CCl_4, water–CS_2 interfaces)?
- How does sp^2 hybridization influence the IFT if choosing benzene–water interfaces over cyclohexane–water interfaces?
- Is there any ionic field between the ionic field and sp^2 hybridization?

Chapter 13

Critical Temperature of a Single-Component System

At temperatures above the critical temperature in a one-component system like H_2O, the physical and thermodynamic properties of a fluid with changing temperature and pressure vary smoothly and continuously, and no discontinuities in fluid properties occur. Such fluid dynamics play a critical role in developing cohesive forces (CFs) and frictional forces (FFs), which could create effective friccohesity in such a valuable fluid. Friccohesity talks about continuity of the fluid except when CFs are transforming to kinetic forces (or FFs), leading to dispersive forces in an array or entropic manner. These various algorithms could be proposed subsequently to be incorporated into numerical fluid flow models that describe the circulation of hydrothermal fluids in the crust. Further it could be of academic and industrial interest to also track the surface energy/tension and the viscosity of such fluids in a similar manner without any gap or the transferring of fluid from one model to another. Such measurements are subjected to the survismeter. There has been a critical trend of a decrease in the viscosity on increasing the temperature, but at the same time the surface energy/tension, which could affect the viscosity, is a most challenge and measurable

Survismeter: Fundamentals, Concepts, and Applications
Man Singh and Sunita Singh

ISBN 978-981-4774-70-3 (Hardcover), 978-0-429-02761-1 (eBook)
www.panstanford.com

task that is logically resolved by friccohesity. In light of mutual or complementary subscription of such data the picture becomes clear about the fluid flow dynamics with variable temperature and even understanding of crust science. Thus friccohesity could explain the truncated science of fluid flow obtained on increasing the temperature. Thus, the viscosity of the liquid phase along the vapor + liquid + halite coexistence curve predicted by the present model is consistent with the known (or expected) relationship between viscosity, temperature, and salinity and shows a decrease in the viscosity with an increase in the temperature, followed by an increase in the viscosity at higher temperatures, reflecting the effect of increasing NaCl concentration.

The temperature-viscosity-phase changes seem a most fascinating science of liquid mixtures that could furnish valuable data [3]. For example, temperature-pressure-mass transfer has a close physicochemical link with chemical thermodynamics in the context of the dynamic viscosity and dynamic surface tension attained through entropic sustainability. The source of the data with respect to the temperature-pressure should be clearly mentioned so that the viscosity data at high temperature are explained. However, it is difficult to measure the viscosity data of mixtures at higher temperatures due to the uncontrolled collisions and nonlaminar fluid dynamics. The survismeter has resolved such restrictions.

13.1 Heat Transfer and Physicochemical Properties

Interestingly, mathematical correlations of physicochemical properties responsible for oil and gas production in wellbore are urgently needed. The most relevant fluid dynamics within the capillary flow could give some integrated look. Study of physical variables in the context of properties like density, heat flow, viscosity, and friccohesity, which deals with CFs and FFs, are being operated within the wellbore. These data furnish an integrated look for a better correlation factor of their interrelationship. Among all such parameters, friccohesity gives a balanced look for modelling of the variables that efficiently work among the structures of the solute, the additive, and the solvent. Specifically, for a process like wellbore during oil and gas production, scientists use an empirical correlations or mechanistic approach algorithm. So typical research in this field is concentrated on a single property analysis, such as that of heat transfer, pressure, or temperature. Here the most proper correlations, regarding the subject, are presented. In light of the above an advanced view for an illustration of oil and gas production is extracted by interrelationship factors of CFs and FFs, which are obtained on their mutual rationalized transformation, noted as friccohesity. It is a new physicochemical super intermolecular forces (IMFs) model to predict molecular structure–driven location to separate or to bind the gas with oil.

The solution describes the wellbore's physical properties, at every depth and at every stage of production, even for the single whole oilfield. For such situations, the friccohesity model could be a novel one for finding a solution in a most strategically controlled way that predicts continuity in the localized molecular domain. It is essentially required for the transfer of heat, pressure, and temperature to transform CFs (potential energy) into FFs (kinetic energy) in a most uniform manner. The interfacial tension (IFT) is an essential property of such models. It finds rigidity of the molecules and their CFs and FFs required for shifting from their equilibrated structural position to a nonreversible position. The temperature model is based on the pressure model, which could not be treated in isolation but depends on density, structural makeup, functional groups, and electrostatic forces. Hence the synchronized studies project an integrated view in regard to interconnected physicochemical parameters. The novel ideas for considering the basic theories of the mutual solubility of components in light of the hydrocarbons liquid–gas interfaces are of use.

Production rate, pressure, temperature, density, solubility, compressibility, formation volume factor, viscosity, heat capacity, surface tension, gas/oil ratio, gas/water ratio, mass rate, bubble point pressure, flow direction, flow regimes, velocity, hydrostatic and kinetic pressure drop, and Reynolds number are considered. A pressure loss across the wellbore during production contains FFs, but how the FFs are generated is not mentioned. Therefore, here friccohesity becomes an inevitable factor that incorporates the CF as well the FF. Density and viscosity are the most important flow parameters of an actual fluid sample taken from the field, and it explains the local equilibrium thermodynamics and motions to facilitate flow and mass transfer. Heat capacity of each phase is a function of temperature and density, usually estimated from the experimental data. So intermixing could also be correlated. The temperature and pressure affect density and other physical properties interdependently, considering a complete set of the intramolecular multiple forces as a useful correlation to study well dynamics with respect to the oil and gas relationship. The studies of friccohesity are responsible for developing the oil and gas systems as a heterogeneous thermodynamic model to explain the pressure and temperature variables. These are linked with CFs (potential energy) as well as FFs (kinetic energy), which show an interesting simulation on fluid dynamics of oil and gas in combination forming a heterogeneous thermodynamic system. Such studies explain connivance of physicochemical properties and interacting forces.

13.2 Micelle with Effective Membrane Strength

Contrary to the above science the Lennard-Jones potential (LJP) becomes most effective in dealing with the molecular activities noted in Fig. 13.1. The like r_e as the equilibrium distance between the hydrophilic and hydrophilic and the hydrophobic and hydrophobic parts develops the most workable micelles. Such micelles with effective membrane strength can hold larger amounts of the drug, air, dye, and others (Figs. 13.2–13.5).

The survismeter capillary response with the states and structures of the hydrophilic and hydrophobic surfaces could be of interest, which are noted in Figs. 13.2–13.5.

Since the micelle membranes have both negative and positive poles of water, if the micelles are made in aqueous mixtures, these poles exert opposite forces to bring the micelles together with definite attractive forces. Similarly, inside the micelles the

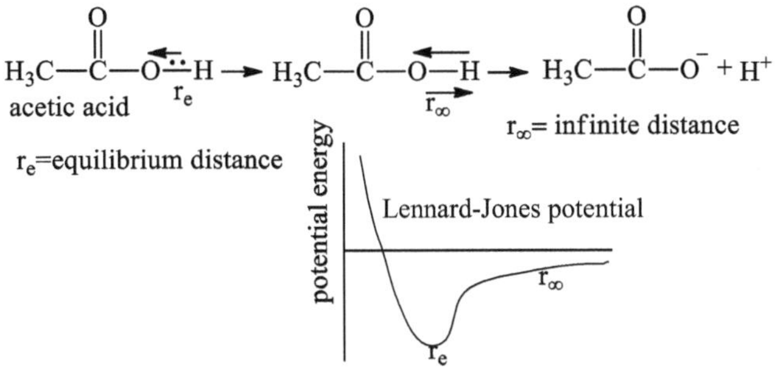

Figure 13.1 Lennard-Jones potential applicable for interactions as a function of distance.

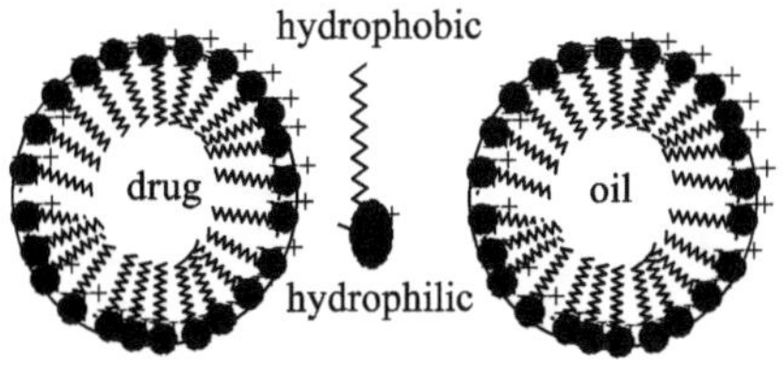

Figure 13.2 Drug or oil causing hydrophobic and hydrophobic interactions with alkyl chain of surfactants in a small population of normal micelles.

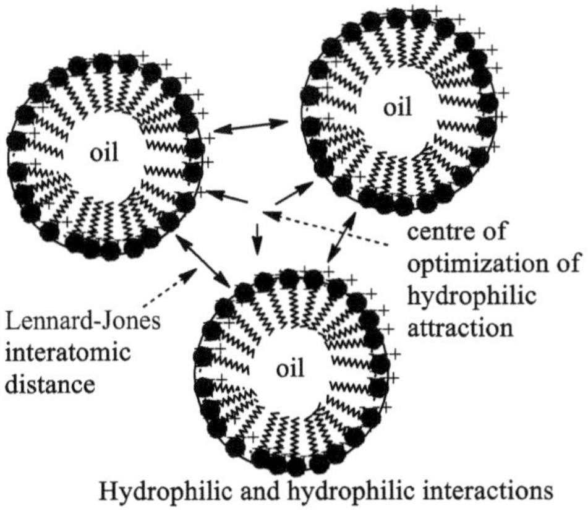

Figure 13.3 Larger micellar population causing functional sites with kinetically water molecules developing hydrophilic and hydrophilic interactions.

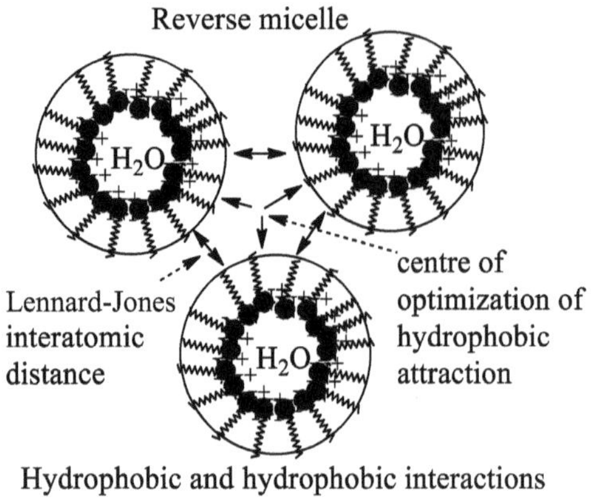

Figure 13.4 Reverse micelles with hydrophobic and hydrophobic interactions.

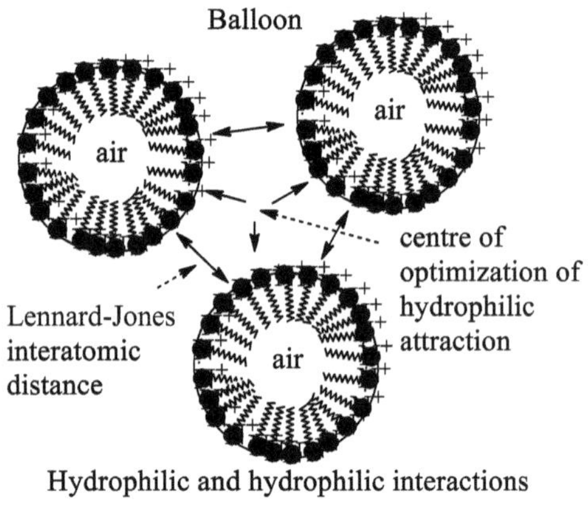

Figure 13.5 Mechanism of balloon formation on air entry into core of micelles via thin film made out of hydrophilic and hydrophilic interactions.

hydrophobic and hydrophobic forces of attractions work, which leads to stabilizing the micelles.

At the center of the micelle optimization occurs, which could develop the maximum charge density with stronger structures of water if the charge density is more aligned. This is the reason that the survismeter measures the charge density on particles in a way similar to taking the zeta potential measurements. Thus, the number of salt or surfactant molecules and temperature could determine the zeta potential of the sample. The amounts of water trapped by the reverse micelles (oil or air in the case of normal micelles) influence the fluid dynamics.

The stability of the micelles is controlled by the hydrophobic forces of the alkyl chain of the surfactant and also of the oil or the dug that approaches the core of the micelles for stabilization. If in place of the oil or drug, air also approaches the core, then the air gets expanded largely in the absence of its binding forces with itself, so a bubble is formed. The bubble forces apply larger internal pressure and nanosized thin films of water are formed due to water and water forces caused due the dipolar interactions of the water at the periphery of the micelles.

***Question*:** Study the size of the bubble with respect to the size of the alkyl chain of the surfactants.

Chapter 14

Friccohesity of Fortified Beverages: Excellent Physicochemical Properties

The ordinary drinkable water does not have vitamins, minerals, fiber, antioxidants, or slenderizing properties. Thereby vitamins and others are infused in the water, which is known as vitamin-infused water. Such water samples do have the most interesting friccohesity because of the controlled interaction of the particles or the vitamins that have been infused. In general, the water products that increase energy contain high levels of caffeine or more stimulants than the most popular soft drinks. In some cases fortified water includes plant products like guarana, which contains a good amount of caffeine, in their formulations. Similarly, an extract of mint, garlic, fenugreek, coriander, and ginger could be infused in water. Water-soluble vitamins are passed out of the body through your urine, so in such cases, their larger intake is not justified, it being merely a waste of money; but fat-soluble vitamins A, D, E, and K are stored in the body, which may lead to an overdose that could cause some negative health effects. The structure of vitamin A (Fig. 14.1) could be infused with water, and their comparative data could be measured.

Therefore, there is an urgent need to infuse or fortify water with vitamins and other energetics in a most controlled manner so that their bioavailability is enhanced and they do not go waste with urine.

Survismeter: Fundamentals, Concepts, and Applications
Man Singh and Sunita Singh

ISBN 978-981-4774-70-3 (Hardcover), 978-0-429-02761-1 (eBook)
www.panstanford.com

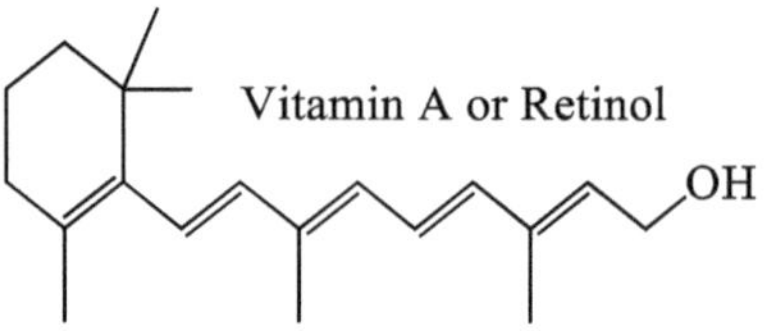

Figure 14.1 Structure of vitamin A.

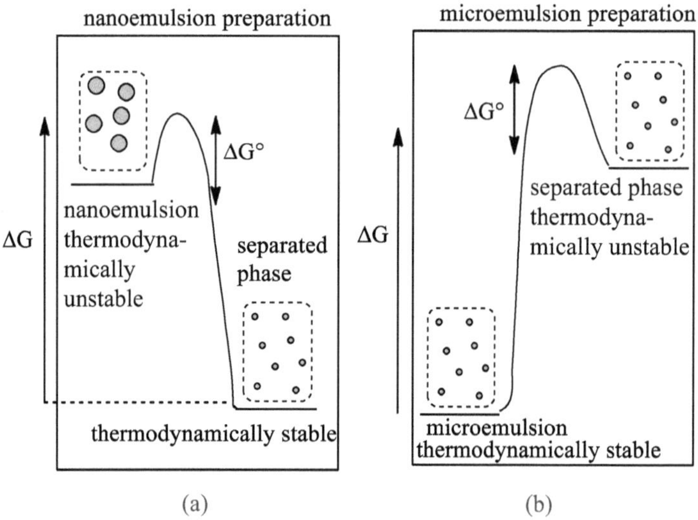

Figure 14.2 Thermodynamics of nanoemulsions and microemulsions.

The vitamins break down over time and are not available in as much amounts as noted on their labels for intake. Therefore, a proper thermodynamic understanding of a mechanism to infill the vitamin, fiber, minerals, and other energetics is a prerequisite condition and explained in Fig. 14.2(a, b).

14.1 Incentives and Understanding of Food Intake

Question: Determine the bound amount of the infused vitamin and others using the survismeter.

Checking Saturated Nanoemulsion Formation

Saturated nanoemulsions or microemulsions do not have free water because the added water in the formulation is fully utilized in developing nanosized thin layers around the oil or vice versa. Therefore, in such a situation anhydrous $CuSO_4$ may not attract the water and no color change occurs without the formation of $CuSO_4 \cdot 5H_2O$ hydrous form. Had there been free water available then such tests could have given a blue color of $CuSO_4 \cdot 5H_2O$. It depicts the presence of water or free water in microemulsions or nanoemulsions. The test checks the strength of the microemulsion. Similarly, such a device or approach could determine the affinity of the water with a series of ethanol, salts, amino acids, and proteins. For example, acetic acid and water both are mixed together and then salts like NaCl, KCl, and $CaCl_2$ could be added till the phase is separated. Therefore, the friccohesity of such systems could be the most interesting data.

14.2 Friccohesity and SAR

The nature and constituents of the molecules affect friccohesity data that illustrate and elucidate the internal molecular mechanism. For example, the following systems are noted:

- Salts that develop ionic interaction, inducing ionic hydration
- Organic molecules that increase friccohesity
  - Surfactants and zwitterionic structure
  - Dendrimer, flavonoids, surfactants, and Pt complexes
  - Starch and aggregates of coalescence

Chapter 15

Structural Interactions with Viscosity Coefficients

15.1 Interionic Interaction Depicted by Viscosity A-Coefficient

Fundamentally, viscosity A- and B-coefficients are determined by the following equation:

$$\frac{\eta}{\eta_0} = \eta_r = 1 + Am^{1/2} + Bm + Dm^{3/2} \quad \text{or} \left(\frac{\eta_r - 1}{m^{1/2}}\right) = A + Bm^{1/2} + Dm$$

Since the nominator is unitless and the denominator has molality unit the unit of viscosity A-coefficient is:

$$\left(\frac{\eta_r - 1}{m^{1/2}}\right) = A \quad \text{or} \quad \frac{1}{\left(\frac{\text{mol}}{\text{cm}^3}\right)^{1/2}} = A \quad \text{or} \quad \frac{\text{cm}^{3/2}}{\text{mol}^{1/2}} = A$$

$$\text{or} \quad \text{A} = \text{cm}^{3/2}\text{mol}^{-1/2}$$

Survismeter: Fundamentals, Concepts, and Applications
Man Singh and Sunita Singh

ISBN 978-981-4774-70-3 (Hardcover), 978-0-429-02761-1 (eBook)
www.panstanford.com

15.2 Intermolecular Interaction Depicted by Viscosity B-Coefficient

Similarly, the unit of B-coefficient in the case of molecular interactions is noted as:

$$\frac{\eta}{\eta_0} = \eta_r = 1 + Bm + Dm^2 \quad \text{or} \left(\frac{\eta_r - 1}{m}\right) = B + Dm$$

$$\left(\frac{\eta_r - 1}{m}\right) = B \quad \text{or} \quad \frac{1}{\left(\frac{\text{mol}}{\text{cm}^3}\right)} = B \quad \text{or} \quad \frac{\text{cm}^3}{\text{mol}} = B$$

$$\text{or} \quad \text{B} = \text{cm}^3\text{mol} \quad \text{or} \quad \text{B} = \text{dL mol}^{-1}$$

15.3 Structure Breaking and Making Interactions with Viscosity A- and B-Coefficients

For viscosity determination for aqueous (1.0 to 4.5 mol kg^{-1}) NaCl solutions at desired are measured. The viscosity of water as a solvent is 0.8903 mPa.s, which is noted as reference or standard data. If the viscosity of a solution is increased on adding salts, the B or $[\eta]$ is positive and the salt is the structure maker, and if the same is decreased, then the solute is the structure maker. The details are given in Figs. 15.1 and 15.4, where the negative slope indicates a structure breaker whereas the positive slope indicates the structure maker.

The ionic solvation of the ions is depicted in Fig. 15.2. It causes a positive charge on the outer surface of the cationic solvation sphere and a negative charge on the anionic solvation sphere. Such alignment of opposite charges of solvation spheres leads to strengthening of the cohesive force (CF), with a higher surface tension as well as viscosity.

15.4 New Equation and Physicochemical Indicators

Table 15.1 explains the nature of liquid mixtures with physicochemical indicators.

Table 15.1 Nature of liquid mixtures with physicochemical indicators

Surface tension	Viscosity	Density	Friccohesity	Conductance	Nature
Higher	Lower	Lower	Lower	Higher	Hydrophilic
Lower	Higher	Higher	Higher	Lower	Hydrophobic
Higher	Higher	Higher	Moderate		Ionic

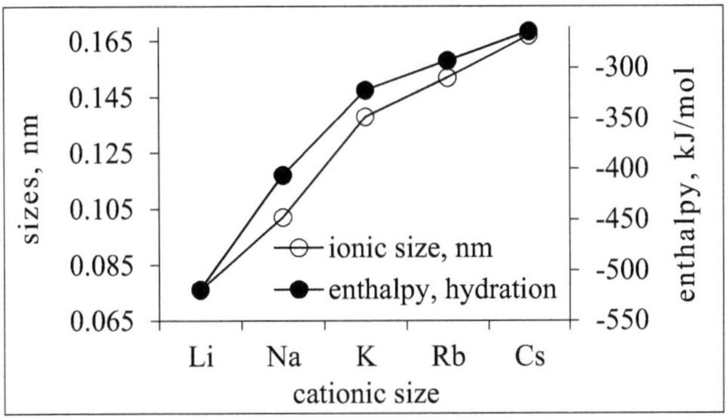

Figure 15.1 Relationship between the size of cations of alkali metals and their enthalpy in aqueous solutions, their chloride salts were used for enthalpy determination.

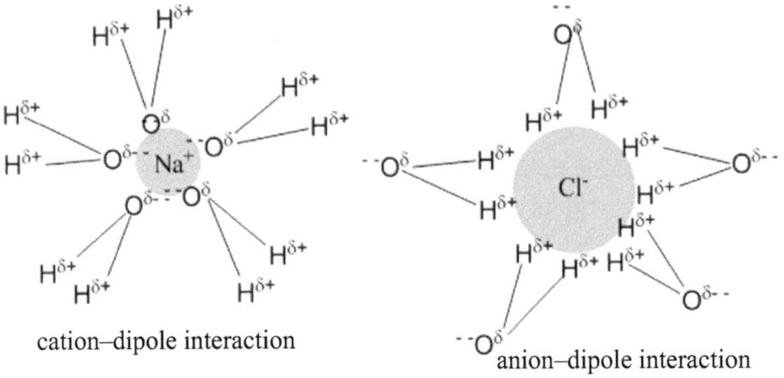

Figure 15.2 Hydration spheres of cation and anion with opposite charges.

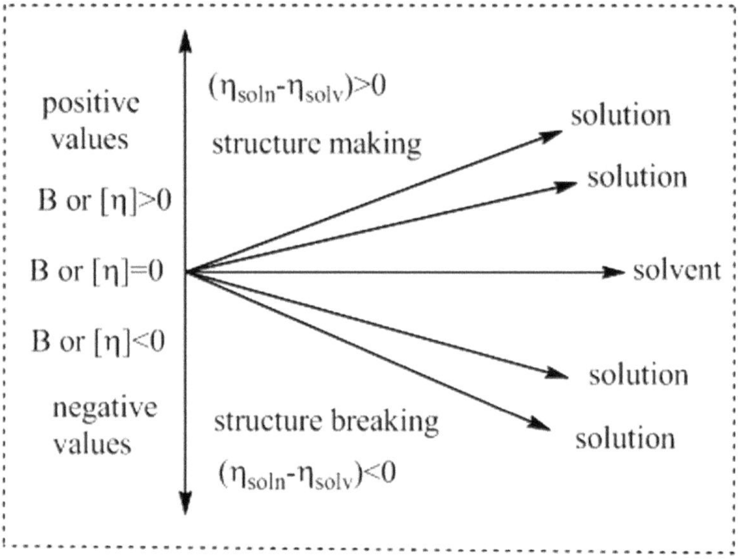

Figure 15.3 Structure making and breaking interaction mechanisms depicted by the positive and negative intrinsic viscosities, respectively.

The viscosity (Fig. 15.3) could increase or decrease as per interactions of the solvents. For example, for stronger interactions on structure breaking, the slope is positive, and for weaker interactions of structure making, the slope is negative. Surface tension of a solute (NaCl) in water is noted in Fig. 15.4.

15.5 Structure Maker and Breaker Theories

The interactions that disrupt the hydrogen bonding (HB) of water are noted as structure breaking, and the solute that strengthens the structures is explained as a structure maker, noted as under:

$$(\eta_{\text{solution}} - \eta_{\text{solvent}}) > 1 = \text{structure maker as } \left(\frac{\eta_{\text{r}} - 1}{m}\right) > 0$$

$$(\eta_{\text{solution}} - \eta_{\text{solvent}}) < 1 = \text{structure maker as } \left(\frac{\eta_{\text{r}} - 1}{m}\right) < 0$$

These equations indicate a disruption of HB. In the first equation the hydrogen bonds of the solvent are strengthened, whereas the

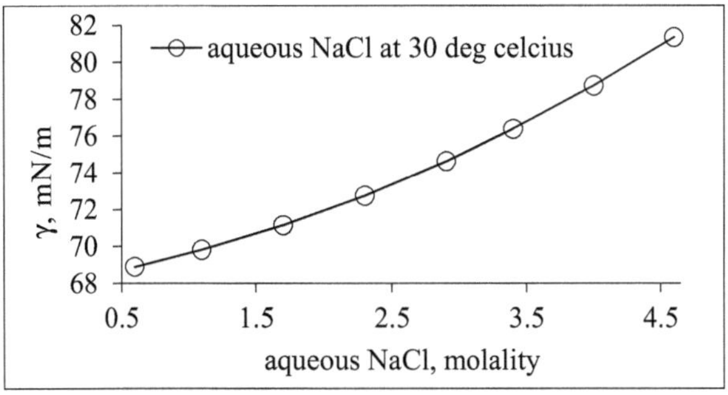

Figure 15.4 Increasing surface tension values of aqueous NaCl solution with increasing concentrations of the NaCl to determine the physicochemiconics of aqueous electrolyte solution.

second equation depicts a weakening of the HB so that a structure is not regained. These concepts are applicable in the case of the salt solutions, but in the case of an emulsion, the situation differs because in such solutions, the structure of the water surrounds the nanoparticle (NP). A separate mechanism is applied for salts and surfactants. In the case of salts, ionic hydration around ions occurs with a stronger ionic field, which strengthens the CF between newly aligned water molecules around oppositely charged ions. Such a mechanism increases both the surface tension and the viscosity due to a stronger CF. In bulk the ionic fields of cations and anions both work effectively. So despite the embedded ions the ionic fields become operational and differ from the Schrödinger potential box. The surfactants develop a different interacting mechanism with water because their one part is hydrophilic and other is hydrophobic in nature. So the ionic hydrations of cations and anions contribute equally to the development of the CF despite the presence of cations in an embedded state. The hydrophilic part weakly initiates the ionic interactions with water and alkyl chain with equal charge distribution and are in a position to alter or reorient the structured water. Both parts are not in a position to contribute their equal share to cause the CF. The surfactant, due to less accommodation in the bulk water, gets realigned near the air–water interface.

$$\left(\frac{\eta_r - 1}{m}\right) > 0$$

Chapter 16

Detection Limit of the Survismeter

A detection limit is estimated from the mean of blank, standard deviation of blank, and some confidence factor. Several detection limits are used, like instrument detection limit (IDL), method detection limit (MDL), practical quantification limit (PQL), and limit of quantification (LOQ), which could be determined using the following standard equations:

$$\sigma = \sqrt{\frac{\Sigma\,(\eta - \bar{\eta})^2}{n-1}} \quad \text{or} \quad \text{SE} = \frac{\sigma}{\sqrt{n-1}} \quad \text{and} \quad \bar{\eta} = \sum_{1}^{n} \frac{\eta}{n}$$

$$\text{Lower limit} = \bar{\eta} - 1.96\frac{\sigma}{\sqrt{n}} \text{ and higher limit} = \bar{\eta} + 1.96\frac{\sigma}{\sqrt{n}}.$$

Fractional uncertainty of uncertainty ($\Delta\sigma$) is noted as the fractional standard deviation of standard deviation (σ), which depends on a number of measurements, noted as under:

$$\text{Fractional uncertainty } \frac{1}{\sqrt{2(n-1)}} \text{ and uncertainty} = \frac{\sigma}{\sqrt{n}}$$

Total uncertainty in $\eta = f(m, T)$.

$$\delta\eta = \sqrt{\left(\frac{\partial\eta}{\partial m}\delta m\right)^2 + \left(\frac{\partial\eta}{\partial T}\delta T\right)^2}$$

Survismeter: Fundamentals, Concepts, and Applications
Man Singh and Sunita Singh

ISBN 978-981-4774-70-3 (Hardcover), 978-0-429-02761-1 (eBook)
www.panstanford.com

Relative uncertainty

$$\frac{\delta\eta}{\eta} = \sqrt{\left(n_m \frac{\delta m}{m}\right)^2 + \left(n_T \frac{\delta T}{T}\right)^2}.$$

The experimental error $\delta\eta$ is determined as

$$\delta\eta = \bar{\eta} - \eta_{\text{std}}.$$

$$\%\ \text{error} = \frac{(\bar{E} - A)}{A} \times 100 \text{ and } \eta = \bar{\eta} \pm \sigma.$$

$$\frac{\delta\sigma}{\sigma} = \sqrt{\left(n_m \frac{\delta m}{m}\right)^2 + \left(n_T \frac{\delta T}{T}\right)^2 + \left(n_{\text{vft}} \frac{\partial \text{vft}}{\text{vft}}\right) + \left(n_{\text{pdn}} \frac{\partial \text{pdn}}{\text{pdn}}\right)}.$$

An uncertainty is determined as under:

$$\frac{\delta\sigma}{\sigma} = \sqrt{\left(\frac{\delta m}{m}\right)^2 + \left(\frac{\delta T}{T}\right)^2 + \left(\frac{\partial \text{vft}}{\text{vft}}\right) + \left(\frac{\partial \text{pdn}}{\text{pdn}}\right)}$$

Friccohesity motions are calculated as under:

$$\eta = Ae^{-\frac{E}{\text{pdn}}} \text{ or } \ln\eta = \ln A - \frac{E}{\text{pdn}}$$

An increase in temperature decreases viscosity, and if the pendant drop number (pdn) increases, the viscosity decreases. The pdn increases when the cohesive force (CF) decreases on stronger dispersion in the case of an emulsion or a nanoemulsion. So the oil model of friccohesity could be applied here. In the oil model the CFs of oil and water remain constant but surfactant addition increases the pdn.

16.1 Entropy Determination from Activation Energy

The equation for the first law of thermodynamics is noted as:

$$dq = dE + PdV - \sum_i \mu_i dn_i \tag{16.1}$$

Here, dq is the heat content, dE is the internal energy, P is the pressure, dV is the volume change, μ_i is the chemical potential of species i, and dn_i is the number of moles participating in the process. There is no volume change if the solution is of a fixed composition ($dn_i = 0$) and $dV = 0$. As the process is covered under the noncovalent mechanism no redox activity occurs. Equation 16.1 is rearranged as Eq. 16.2.

$$dq = dE \tag{16.2}$$

But $dq = TdS$, so putting the value of dq in Eq. 16.2 gives a new equation, Eq. 16.3.

$$TdS = dE \quad \text{or} \quad dS = \left(\frac{dE}{T}\right)_{p,n_i} \tag{16.3}$$

Here, dE could be taken as activation energy determined by viscosity. Therefore, the entropy data could be directly calculated with activation energy.

16.2 Effect of Particle Size on Physicochemical Parameters

Figure 16.1 depicts the size-based dynamics of the nanoparticles (NPs) in nanoemulsions. Several processes are controlled by the sizes of particles involved in processes like dynamic light scattering (DLS), gel permeation chromatography (GPC), ultraviolet spectroscopy. Currently, friccohesity and tentropy both have taken a center stage to further redefine and redesign the sciences and working mechanisms of physicochemical sciences through a

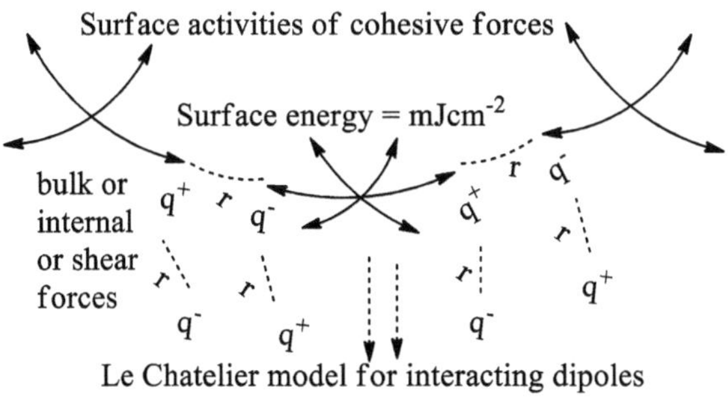

Figure 16.1 Surface and shear forces contribute in fluid dynamics in forming micelles.

rationalized contribution of localized activities of molecules. Such approaches create an activity in the atoms, bonds, electrostatic poles, pi bonds, and sigma bonds to generate local entropy, noted as tentropy. So the larger the particle size, the higher is the viscosity; and the lower the surface tension, the higher is the friccohesity. The larger-sized particles contain a larger inner defunct area with no change in potential energy ($\Delta\mu = 0$), or higher CFs produce a higher surface tension and lower friccohesity. Lower friccohesity (σ) engages a lesser amount of the solvent with the particles. The lower value of friccohesity infers a weaker interacting ability of the molecules or capping of the NPs. Thus the lower the friccohesity value, the weaker is the interacting ability or the activity (a) of the particles or the molecules, which is formulated as under:

$$\sigma \propto a \quad \text{or} \quad \sigma = k \cdot a \tag{16.4}$$

The k here is the interacting indicator. The a activity of the surfactant is determined as

$$a = e^{\frac{\mu_i - \mu_i^0}{RT}} \quad \text{or} \quad \ln a = \frac{\mu_i - \mu_i^0}{RT} \quad \text{or} \quad \mu_i = \mu_i^0 + RT \, \ln a. \tag{16.5}$$

The particles get activated and start partitioning from one level to another as per their activity energy. Z is the canonical partition function, noted as under:

$$Z = \sum_i e^{-\frac{E}{k_B T}} \tag{16.6}$$

In Eq. 16.5, μ_i and μ_i^0 are chemical potentials of i species in solution and in a standard state, R is the gas constant, and T is the temperature. At the same time the smaller particles utilize more solvent with less inner defunct area, utilizing its higher chemical potential. The larger-sized particles show higher surface energy or tension with lower viscosity and higher conductance with higher heat transfer. The smaller the particle size, the larger is the volume engagement by the particles, which is expressed as under:

$$\text{particle size} \propto \frac{1}{\text{engaging solvent volume}} \tag{16.7}$$

The smaller-sized particles are closely packed, with less fractional volume, higher viscosity and friccohesity, and lower surface tension and lower conductance. The systems become congested with a

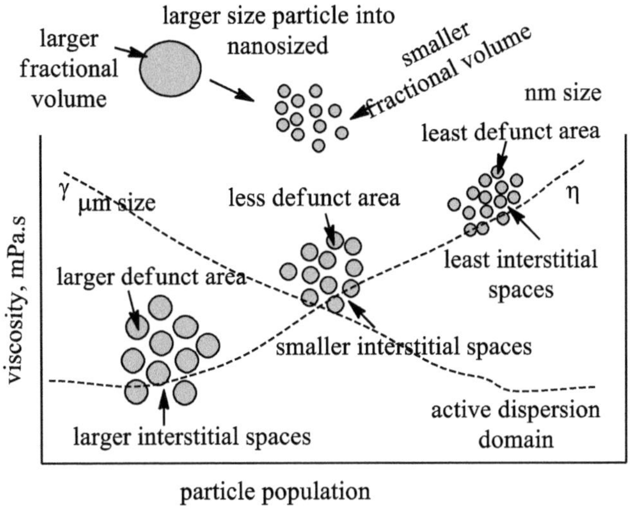

Figure 16.2 Decreasing particle sizes of emulsions with increasing viscosity and decreasing surface tension.

larger number of smaller-sized particles (Fig. 16.2). Therefore, the particle sizes and friccohesity data offer an advanced understanding of the materials' interactions for developing newer applications. Such sciences could develop efficient soaps and dyeing processes and solar radiation trapping activities as well as release and flow electrons in liquid medium. Thus, the state of materials and their properties are intimately correlated. For example, for larger-sized particles the surface tension is lower and the viscosity is higher, along with higher friccohesity values. Nanometer-sized particles get fully mixed with the medium because they come under a range where the atomic forces of the medium and of the NPs come from the Lennard-Jones potential (LJP), which are strongly bound and remain stable.

The UV scattering, CF, contact angle, friccohesity, molecular distribution depicted by viscosity, activation energy, and hydrodynamic radii are mutually connected, so the UV–Vis absorption depicted by A could be correlated as under:

$$A = \log_{10}\frac{I_0}{I_e} = \varepsilon lc,\ \log 10 = 1 \text{ so } A = \frac{I_0}{I_e}$$

If $I_0 = I_e$, $A = 1$; such a situation is developed in the case of ideal liquid mixtures. But for real solutions $A = I_0/I_e > 1$. Experimental measurements are made in terms of transmittance (T), which is defined as $T = I/I_0$, where I is the light intensity after it passes through the sample and I_0 is the initial light intensity. The relation between both A and T is as under:

$$\frac{1}{T} = A = \log_{10}\frac{I_0}{I_e} = \varepsilon lc, \log 10 = 1 \text{ so } A = \frac{I_0}{I_\text{e}}$$

$$x_2 \approx v \cdot m \cdot M_1$$

The size of nanoemulsions also correlates with the response of the UV light as interactions between light and matter are also controlled by the size of the light photon and of the molecule or NPs like TiO_2, ZnO, CdS, Au, and Ag.

Chapter 17

Particle Size Distribution and Particle Packing

The magnitudes of the surface tension, viscosity, conductance, and friccohesity infer the size and interacting activities of the particles. For example, a polydisperse population with a broad size distribution packs more closely than a monodisperse sample. The effects on viscosity are explained with reference to the Krieger–Dougherty equation. The equation predicts an increase in the medium's viscosity when particles are added. An increase depends on the concentration of particles.

$$\frac{\eta}{\eta_0} = \eta_r = \left(1 - \frac{\varphi}{\varphi_m}\right)^{-[\eta]\varphi_m} \quad \text{or} \quad \eta_r = \left(1 - \frac{\varphi}{\varphi_m}\right)^{-B\varphi_m}$$

The $[\eta]$ or B intrinsic viscosity equals 2.5 for a sphere, φ is the volume concentration of particles, φ_m is the maximum packing, η is the viscosity of the suspension, and η_0 is the viscosity of the medium. So η_r is the relative viscosity and is noted as under:

$$\eta_r = \left(1 - \frac{\varphi}{\varphi_m}\right)^{-2.5\varphi_m}$$

If viscosities of the cement paste and the concentration of aggregates are known and the maximum packing of the particles is determined,

Survismeter: Fundamentals, Concepts, and Applications
Man Singh and Sunita Singh

ISBN 978-981-4774-70-3 (Hardcover), 978-0-429-02761-1 (eBook)
www.panstanford.com

then the viscosity of the concrete is calculated. For example, Struble used a coaxial rheometer with a gap between cylinders of 0.7 mm. This gap is not as small as a mean gap between the aggregates in concrete (0.12 mm to 0.26 mm). The rheological properties of concrete are measured using a rheometer. Conductance with respect to existing restrictions of the solution is connected to the other parameters as:

$$X \propto \left[\left(\frac{\eta}{\gamma}\right)\left(\frac{\pi}{\lambda}\right)\right] \quad \text{Or} \quad X = k\left[\left(\frac{\eta}{\gamma}\right)\left(\frac{\pi}{\lambda}\right)\right]$$

Here, X is the proposed physicochemical property and η, γ, π, and λ are viscosity, surface tension, osmotic pressure, and conductance of the solution sample, respectively. The bonding and uniform particle distribution weaken both the cohesive force (CF) in the intermolecular force (IMF). The size depends on the parameters explained in the following equation:

$$\text{size} \propto [(\text{CF})(\text{C}_n)(\mu)(f_{\text{group}})] \quad \text{or} \quad \text{size} = k[(\text{CF})(\text{C}_n)(\mu)(f_{\text{group}})]$$

Here, CF is for cohesive force, C_n is the carbon number in the alkyl chain of molecules like the surfactant, μ is the chemical potential, and f_{group} is the functional group. For a monodisperse sample a maximum volume fraction is ~62%. With a polydisperse sample the smaller-sized particles could fill the gaps between the larger ones and a maximum volume fraction is ~74%. An increase in the particle size distribution (PSD) for any given volume fraction of solids reduces the viscosity of the system. The PSD could be a valuable tool for manipulating the viscosity and friccohesity both. It has a fixed volume fraction. For example, viscosity as a function of fraction of the large or small graphene particles with an epoxy could lead to strengthening the stability of the resultant material. It creates a synergistic effect when particles of both sizes are present at certain concentrations that give a lower viscosity than that achieved using a monodisperse sample of either size. These values depict a significance of the PSD that is expressed by viscosity. On loading the polymer or the large-sized solids, the viscosity could be kept same on broadening the PSD. The viscosity is increased by using particles with a narrower size distribution. Such loading causes an additional pressure, which is dealt with using a new approach, given as under:

17.1 Pressure Interconversion Units

The pressure applied in our day-to-day applications is expressed as pascal, kilopascal, and megapascal units. These units are converted into other units as under:

$$1 \text{ Pa} = 1 \text{ Nm}^{-2} \quad \text{or} \quad \frac{1 \text{ N}}{\text{m}^2}, \; 1 \text{ kPa} = 1 \times 10^3 \text{ Pa}, \; 1 \text{ MPa} = 10^{6\,\text{Pa}}.$$

The pressure is related to force in Newton/area, where $F = ma$ and $F = \text{kg·m}^{-2}$, as under:

$$P = \frac{F}{A} \quad \text{or} \quad \frac{\text{mg}}{A} = \frac{\rho V g}{A} = \frac{\rho A\text{hg}}{A} = \rho\text{hg}$$

$\rho^{\circ}_{\text{hg}}C = 13.5951 \text{ g/cm}^3$ and $h = 1$ mm and 1 torr is noted as

$$1 \text{ torr} = \rho\text{hg} = \left(13.5951\frac{g}{\text{cm}^3}\right)(1 \text{ mm})\left(9.8066\frac{\text{m}}{\text{s}^2}\right).$$

The units are $1 \text{ mm} = 10^{-3}$ m and $1 \text{ g·cm}^{-3} = 10^{-3} \text{ kg·}10^6\text{m}^3$; putting these values in the equation:

$$1 \text{ torr} = \rho\text{hg} = (13.5951 \times 10^{-3}\text{kg} \cdot 10^6\text{m}^{-3})(10^{-3}\text{m})\left(9.8066\frac{\text{m}}{\text{s}^2}\right)$$

1 torr is noted as under after putting values and calculating:

$$1 \text{ torr} = \rho\text{hg} = 13.5951 \times \text{kg} \cdot \text{m}^{-1} \times 9.8066\,\text{s}^{-2}$$

$$1 \text{ torr} = \rho\text{hg} = 133.322 \text{ kg m}^{-1}s^{-2}$$

$$1 \text{ Pa} = \frac{F}{A} = \frac{ma}{A} = \frac{\text{kg} \cdot \text{m} \cdot \text{s}^{-2}}{A} = \frac{\text{kg} \cdot \text{m} \cdot \text{s}^{-2}}{\text{m}^{-2}} = \text{kg} \cdot \text{m}^{-1}\text{s}^{-2}$$

1 torr 133.322 $\text{kg·m}^{-1}\text{·s}^{-1} = 133.322$ Pa, 1 torr = 1 Pa, and 1 atm = 760 torr. So

$$1 \text{ atm} = 760 \text{ torr}; \; 1 \text{ atm} = 760 \times 133.322 = 101324.72\,\text{Pa}$$
$$= 1.0132472 \times 10^5 \text{ Pa}.$$

Then 1 atm = 760 torr and 1 bar = 0.986923 atm; so 1 bar = 760 × 0.986923 = 750.062 torr.

17.2 Tentropic Emulsion

A tentropic emulsion originates from an unequal charge distribution among atoms, functional groups, or localized structural interstities that develop various orders of molecular and localized activities. So these activities induce localized entropies within a molecule. The nearby localized areas with various atomic motions, charges, and orders are different. The nearby areas within a molecule with various orders of activities in the nanometer range act as the surrounding. It is similar to the surrounding of the working systems in the case of the entropy. For example, ethanol or glycine causes various stretching frequencies in its Fourier-transform infrared (FTIR) spectrum. The variable frequencies are of different orders within a single molecule. In such macromolecular structures the variable arrays of the variable orders at nanorange are noted as tentropy in the nanometer area rather than in macroscopic bodies. For example, the emulsions are made up of longer alky chains as a hydrophobic tail is attached with a hydrophilic head, and due to stronger Brownian motions higher tentropy is developed. It has a higher kinetic energy, so such microemulsions with specific temperatures could be dynamic and hold heat capacity with higher arrays of motions. Also such emulsions with a definite amount of magnetic nanoparticles (NPs) or dendrimers could enhance their effect. The tentropic nanoemulsion could be good for muscle and tissue therapies with definite localized equilibrium thermodynamics (LET) and localized kinetic energy (LKE). These are made by taking most hydrophobic molecules and jerking in ultrasonic mode through a soft silicon flat tape where fluid flow permits a most suitable therapy. So hydrophobic = higher entropy and hydrophilic = higher enthalpy.

17.3 Microemulsion for Food Sciences

Since ancient times, food or herbal pastes or liquid mixtures are widely accepted because of their homogenous mixing and solubilized ingredients. Food scientists and researchers have been developing new ways to improve nutraceuticals and the nutritional value of processed foods. The nanoemulsion technique can increase the water solubility and absorption and decrease the amount of nutrients required to make foods more nutritious. Currently, nanoemulsions are made via high-pressure homogenization, but ultrasonic homogenization and electrified coaxial jet emulsification are being used in lab settings to enhance stability and absorption of nutraceuticals. Nanoemulsions of food materials are safer in osmotic activities and do not apply much pressure on the cell membrane. Researchers are developing nanoencapsulation techniques using micellar cells for more targeted nutrient delivery.

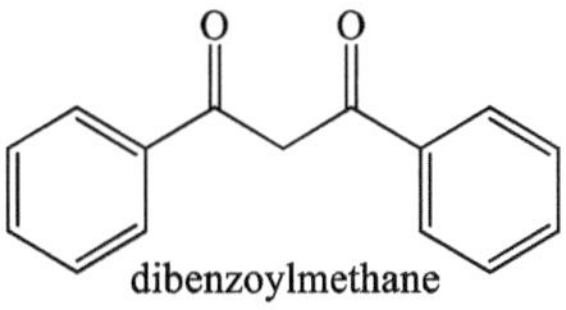

dibenzoylmethane

Homogenizing food contents or ingredients throughout the whole mass of the specific food item has been a challenge to maintain solubility or dispersion in liquid microemulsion. Table 17.1 contains their salient features where the parameters distinguish micro- and nanoemulsions.

Considering Table 17.1, an actual picture of a nanoemulsion and a microemulsion is demonstrated in Fig. 17.1. The binding or potential energy of a single drop (Fig. 17.1a) is higher as compared to that of more drops (Fig. 17.1b and c) because their binding

Table 17.1 Salient features of micro- and nanoemulsions

Microemulsion	Nanoemulsion
Thermodynamically stable = $\Delta G < 0$	Kinetically stable
Long-term stability = $\Delta G < 0$	Short-term stability
Higher surfactant concentration	Lower surfactant concentration
Less expensive nanoemulsion	Expensive
Lower viscosity and friccohesity	Higher viscosity and friccohesity
Higher conductance	Lower conductance
Higher surface tension, less wettability	Lower surface tension, higher wettability
Lower friccohesity	Higher friccohesity

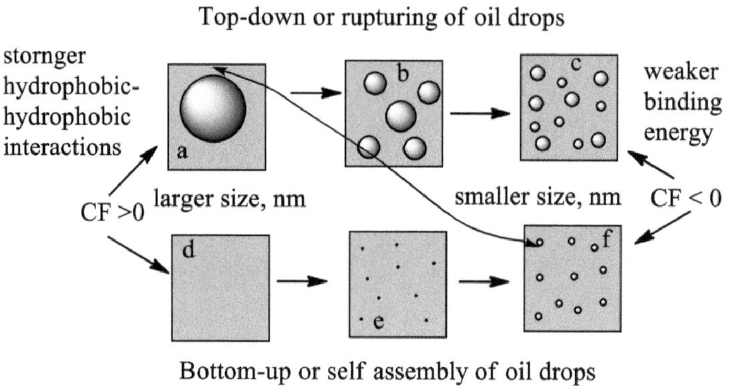

Figure 17.1 Friccohesity: Fundamental science of nanoemulsions.

energy is transformed into kinetic energy. Energy from Fig. 17.1d to Fig. 17.1f is variable, where the hydrophilic interactions are stronger in the case of Fig. 17.1d because the dispersate is fully and structurally soluble rather than dispersing. Such dynamics could align the electrical charges on the molecules to develop an electrical mass transfer.

Figure 17.1 depicts a large-sized drop with higher binding dynamics holding all the oil molecules together with higher stability with stronger hydrophobic-hydrophobic interactions. Surfactants weaken them and release them into the systems (Fig. 17.2). ΔG of Fig. 17.1 is as $a(\Delta G) < b(\Delta G) < c(\Delta G)$. The working structural of cetyl-trimethyl-ammonium bromide (CTAB) is noted as under with philic-phobic parts. Br^- as a counterion develops ionic solvation with a definite ionic field that induces structure breaking activities, contrary to the lowest entropy in the vicinity of the ion and higher

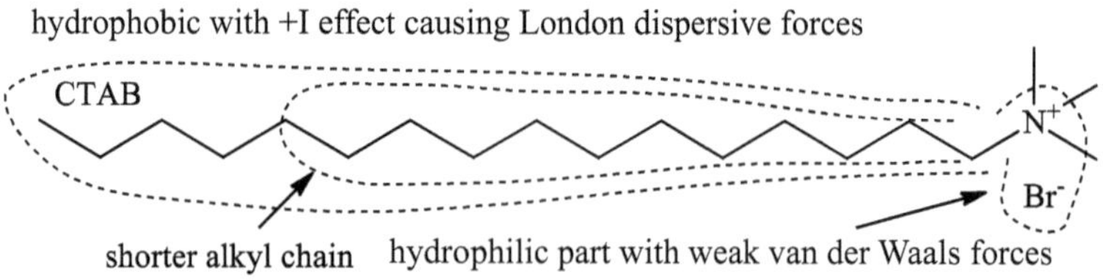

Figure 17.2 Hydrophobicity of cationic surfactant with its increasing alkyl chain along with hydrophilic head.

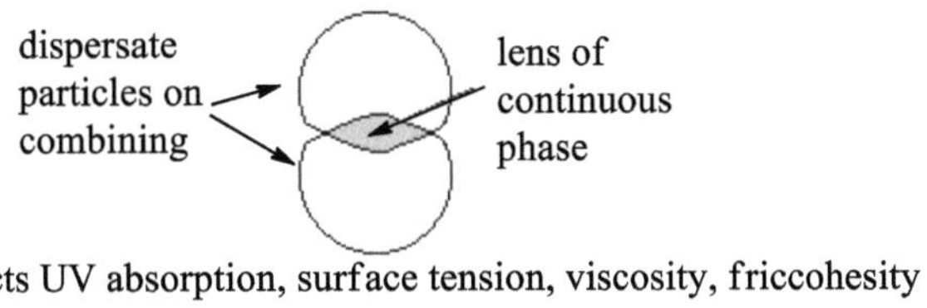

Figure 17.3 Coalescence of nanoparticles in nanoemulsions affecting their physicochemical properties.

entropy outside of the solvation sphere. Such structurally different domain affects the Brownian motions. In such a situation the measurements of friccohesity become relevant in regard to the drug-holding ability of the micelles.

Coalescence is directed related to friccohesity. For example, if the surface tension is lower and viscosity is higher, then droplets are not able to completely coalesce and develop some gaps in between. A dispersion phase is strongly bound, and particles cannot overcome this energy barrier and are unable to coalesce, as shown in Fig. 17.3.

Lower surface tension and higher viscosity favor less coalescence, which is supported by higher friccohesity, lower conductance, and smaller particle sizes. The coalescence is directly connected with friccohesity and an energy barrier of liquid mixtures caused due to stronger solute-solvent interactions. The structural components of the droplets are noted in Fig. 17.4a and b, whereby an

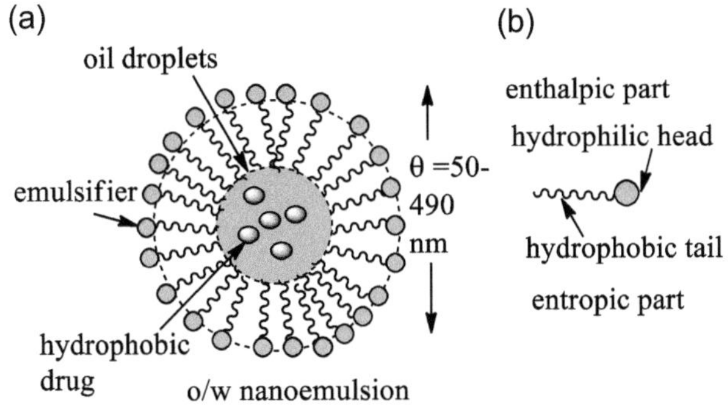

Figure 17.4 Structural dimensions of normal micelles with hydrophobic core attracting hydrophobic oil particles toward the core.

inducing high-power ultrasonic technology could induce stirring ultrasonics through molecular motions. The hydrophobic drug approaches the hydrophobic part at the center in the case of normal micelles. Such systems show a unique set of physicochemical properties (PCPs) as compared to reverse micelle systems. The miceller mixture's activity is different than that of a nonmiceller mixture.

Emulsions do have droplets of single and double layers as per their growth and trapping of polymer molecules if they are added to protect the micelles for longer stability and activity independently depicted in Eq. 17.1.

$$\Delta G = nRT \log \left(\frac{\pi_m}{\tau_m} \right) \tag{17.1}$$

The kinetic state of an emulsion is different as per its layering approach and its dynamic state and motions in mixtures. It is an interesting theory to study the zeta potential and distribution pattern in the case of its polydisperse index (PDI). Figure 17.5(a–c) are given below, which show a covering of the polymer that could act as an excellent medium for a drug carrier or vehicle. Scientists have printed a low-cost radio frequency antenna with graphene ink. Figures 17.5(a–c) explain the different flow patterns that develop a ground to establish a relationship between the size and viscosity

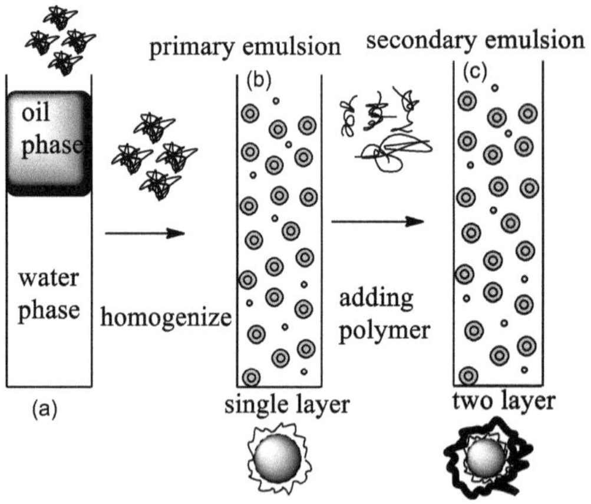

Figure 17.5 Dynamics of nanoparticles of nanoemulsions and their layerwise coating.

of emulsions or solutions. Such kinetic energy patterns could be most effective in predicting the PDI. Thereby, the larger distribution patterns with thousands of structural or reorientational arrays could predict a drastic change in viscosity. Such systems, due to additional kinetic energy, could be used as dynamic soaps, dynamic heat-transferring mediums, electron-aligning mediums, sensors for particle grouping in liquid mixtures, and capping agents and for drug encapsulation by dendrimers. Such a mechanism could also be extended for developing excellent wettability or nano thin films, making for desired light reflection and absorption.

Hydrophilic-based properties bind and sustain stability, but at the same time hydrophobicity keeps particles moving, which is needed for nanoemulsion formation. Each surfactant molecule surrounds the hydrophilic and hydrophobic parts to develop stable interfaces with adequate binding energy based on the intramolecular multiple forces theory (IMMFT). Their mechanism could be supported by a few subprocesses, noted as under:

- Hydrophilic and electrostatic dominations for enthalpic effect
- London dispersive forces (LDF) and friccohesity dominate
- Capping domain and van der Waals forces
- Hydrogen bonding (HB) disruption and functional domains
- pH and temperature domains
- Kinetics domain and thermodynamics
- Combinatorial arrays and destructive arrays

Chapter 18

Unique Experiment

Physicochemical properties investigate structural interactions of salts and surfactants separately in common solvents like water. An integrated fluid dynamics being applied by using survismeter not only distinguishes the interaction mechanism of molecules but also determines their specific effect on solvents structure. Therefore the following newly designed unique experiment is conducted:

***Experiment*:** Determine the surface tension of solvent and solution both under similar conditions.

$$\pi_{sp} = \gamma_{solv} - \gamma_{soln}$$

The $\pi_{sp} < 0$ value indicates the structure maker and the $\pi_{sp} > 0$ value the structure breaker. Aqueous salt solutions behave as structure makers and the surfactants as structure breakers. Ion-solvent interactions are based on the charge center and the length of the alkyl chain of the surfactant. So the longer the alkyl chain, the higher is the surface. So the +I effect becomes the most effective mechanism in getting surface pressure generation. Degree association α_1 depends on mole fractions.

$$\alpha_1 = \frac{[S_1]}{[S_1][S_2]}$$

Both S_1 and S_2 belong to the mole fractions of components 1 and 2. Any resistance created by the void spaces to allow the magnetic

Survismeter: Fundamentals, Concepts, and Applications
Man Singh and Sunita Singh

ISBN 978-981-4774-70-3 (Hardcover), 978-0-429-02761-1 (eBook)
www.panstanford.com

nanoparticles (MNPs) could be created by the boundaries of void spaces. Such boundaries could be the pseudoboundaries.

18.1 Salt Mechanism of Surfactants

Surfactants on dissolving in the dispersing medium get quickly dispersed. The dispersing activity is noted as the degree of dispersing or dissociating degree. Dispersing continues up to a certain limit of surfactant composition, but it gets saturated after a certain composition. So a transition point is attained or established in the dispersing trend, and after a transition point a new dispersion pattern starts. Two slopes exist, the slope value before the transition point and the slope value after the transition point. The degree of dissociation is = the slope after the critical micellar concentration (CMC)/slope before CMC.

$$\text{degree of dissociation } (\propto) = \frac{\text{slope after breaking point}}{\text{slope before breaking point}}$$

The longer the alkyl chain, the lower is the CMC value and the higher are the specific conductance and the friccohesity. A longer alkyl chain occupies larger amounts of the structured water, and no more unoccupied structured water is left to support further dissociation of surfactants.

18.1.1 Calculating Integrals

Calculating an integral of the total volume is permitted for calculating the viscous flow in a capillary.

$$\int_a^b f(x)dx = F(b) - F(a) \quad \text{and}$$

$$\int_0^1 x^{1/2}dx = F(1)^{1/2} - F(0)^{1/2} = 0 - 1 = 1$$

$$\int_0^1 \frac{2}{3}dx = F(1) - F(0),\ \frac{2}{3}(1)^{3/2} - F(0)^{3/2} = \frac{2}{3}$$

$$\text{because} \quad F(x) = \frac{2}{3}x^{3/2}$$

The square root function $f(x) = x^{1/2}$ between 0 and 1 is sufficient to find an antiderivative, that is, a function $F(x)$ whose derivative equals $f(x)$. For example, a solution battery or a dispersion battery or a battery based on ionic change, temperature, pressure, and osmotic pressure works at a differential rate integrated from the initial time to the discharged time.

18.2 Law for Shifting CF to FF in Bulk with Higher Friccohesity

The tension of covalent forces is channelized toward the factors of forces among dissimilar molecules. Such a tendency enhances the viscosity because a similar chemical environment is developed on the solvent molecules getting aligned around the particles or nanoparticles (NPs). This kind of solvent alignment increases the frictional forces (FFs) within the smaller particles or NPs. If the solvent is dipolar, like water, dimethyl sulfoxide (DMSO), or dimethylformamide (DMF), then the solvent's dipoles develop coulombic forces, which increase the surface tension as well as the viscosity. In early 1940s, visbreakers, known as viscosity breakers or known to lower the viscosity of oil, were in use. However, along with viscosity, the friccohesity also explains a perfection and creation of such solvent molecular models. Thus, a shift in covalent forces at the air–liquid interface (ALI) to the bulk phase increases the FFs manifold, which further develop covalent forces on contact. So the law is named as the law of shifting of the cohesive force (CF) to the frictional force (FF) in the bulk phase having a higher viscosity with a lower friccohesity. The reason or factor for such shifts is electronegativity. Figure 18.1a depicts a stronger tension with higher surface energy (J/m^2), which is obtained through the molecular forces factors of the solvent. So a stronger surface energy depicts a quick flow and if the solvent is engaged by solute particles, as is depicted in Fig. 18.2, then it depicts lower surface energy and higher viscosity. So the particles or NPs of the emulsion reorient the solvent molecules around each of the particle. Such arrangements cause stronger friction with a higher viscosity. The fluid dynamics of the emulsion develop interesting

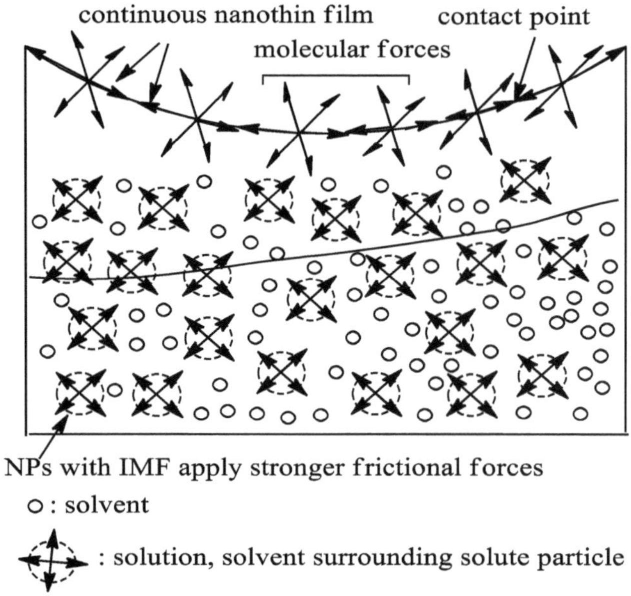

Figure 18.1 Mechanism for origin of surface tension initiated by the unused or vacant molecular force vectors at air–liquid interface forming nanothin film.

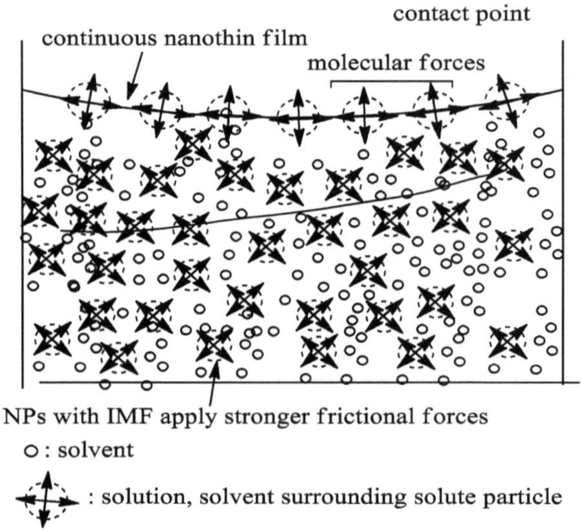

Figure 18.2 Weaker nanothin film at air–liquid interface with stronger interactions with solvent molecules.

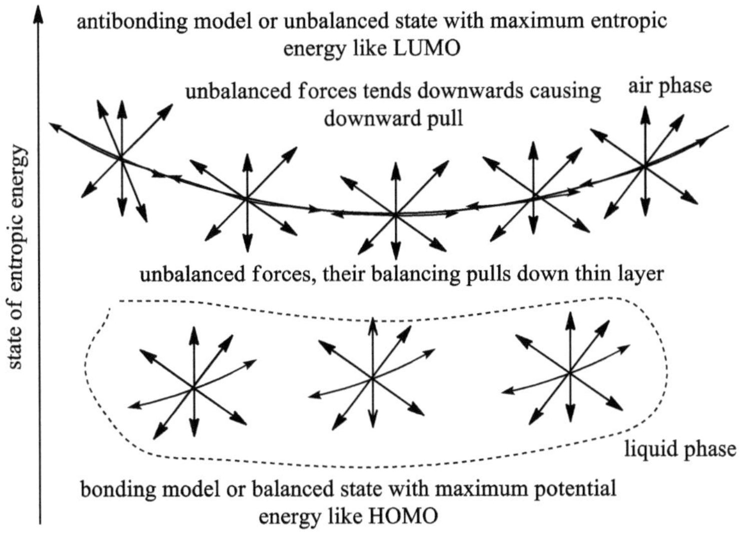

Figure 18.3 Working mechanism of surface tension of a liquid sample where air and liquid phases exist and liquid phase experience a downward pull caused by unbalanced surface forces or surface energy.

data of friccohesity. Figure 18.1 developed lower friccohesity, and the structural state of the emulsion depicted in Fig. 18.2 developed higher friccohesity. The higher friccohesity depicts stronger solvent entanglement of the solute forming the colloids or emulsions. Figure 18.1 depicts a stronger CF with higher tension.

No similar molecules attract or balance these forces. There is a tendency to utilize or nullify these forces. The unbalanced forces are attracted downward as these forces have higher energy and pull the thin layer downward (Fig. 18.3). The natural theory works here because in the bulk that is near the ALI, the molecular force vectors are nullified by the forces of the nearby molecules but the forces that exit at the air interface are not used. Thus, the force vectors that were upward side are not utilized and they are attracted by the lower energy state and in lieu of such activities, a downward pull force is developed. The downward pull causes a tension on the ALI (Fig. 18.3).

A critical domain of friccohesity is the stronger electrostatic forces of molecules that are originated out of the electronegative atoms present in molecules. Similarly, the electron releasing or withdrawing groups could play a contributory role for developing the concept of friccohesity. This theory develops a new concept where stronger cohesive forces (CFs) are generated due to stronger electrostatic forces with higher surface tension of the liquid in the pure state. The viscosity is higher with stronger CFs due to a stronger structural arrangement as a complete homogenous phase. Since the integrated and homogenously developed liquid phases develop a long arrangement of the forces, the single homogenous phase takes time and also acquires larger energy to put the mass of the homogeneous molecules on a flow mode with definite kinetic energy. In such systems there are no canonical partition ensembles or the boxes and the energy are equally distributed and the presence of any kind of heterogeneity with a variety of phases or canonical partitions is ruled out. So a basic difference in these two situations is that the CFs are in the pure state but the FFs are developed due to a distribution of the solute based on solvation or ionic or dispersion, like hydrophobic. Density is not needed to determine such structural changes, so friccohesity is a most effective solution of such liquid mixtures. Thus the advanced science evolved by allowing interactions of more than two molecules in liquid mixtures infer stronger molecular interactions. Friccology is the study of mutual sharing of more than two molecules in liquid mixtures, and both friccology and rheology are mutually interconnected sciences.

Chapter 19

Dynamic and Static Surface Tension

Dynamic surface tension (DST) is the tension per unit length developed at a specified point in a surface and is observed as a function of time. So DST is a part of the nonequilibrium system with $\Delta G < 0$. ΔG is Gibbs energy, which is utilized in gaining equilibrium with time, like coagulations, coalescence, and nanoemulsions. This definition of surface energy introduces a model of interfacial tension (IFT) with adhesive forces at a definite time interval. Static surface tension (SFT), also known as IFT, is the value of surface tension attained in a state of thermodynamic equilibrium that is independent of time. Thus in contrast to DST it refers to a particular surface age like surface tension of water or dimethyl sulfoxide (DMSO) at the air interface. The survismeter measures both DST and SFT at the desired temperature and time.

19.1 Pdn: Most Effective Methodology with the Survismeter for Surface Tension

An expression of the surface is noted in a dilute medium because similar forces are missing in such a medium and the unbalanced forces of liquid sample are not utilized. Thus, pendant drop numbers

Survismeter: Fundamentals, Concepts, and Applications
Man Singh and Sunita Singh

ISBN 978-981-4774-70-3 (Hardcover), 978-0-429-02761-1 (eBook)
www.panstanford.com

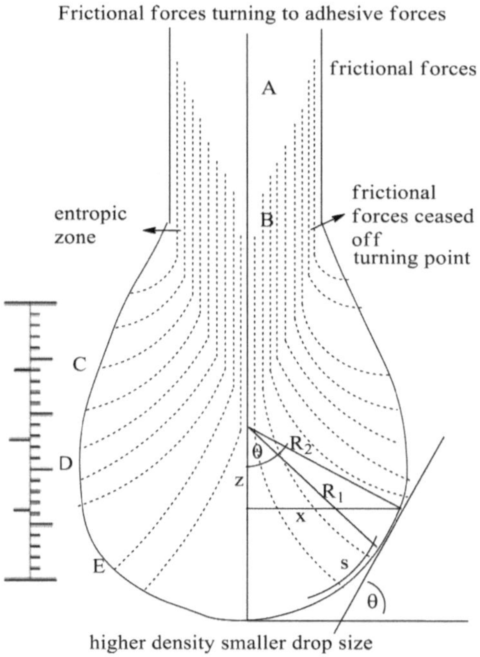

Figure 19.1 Dynamics of cohesive and adhesive forces forming a pendant drop by adhering liquid to the solid surface of lower circumference of hanging capillary.

(pdns) formation with the survismeter is preferred to determine the surface forces or cohesive forces (CFs). The pdn is formed in air as air does not destabilize the CFs and a proper drop is formed naturally. The tip of a solid capillary ending facilitates an expression of the CFs in the formation of the pdn. The pdn indicates a distribution of the CF into a solution formation. The lowest CF causes the lowest surface tension with the highest friccohesity, with the least dipole moment and less zeta potential (Fig. 19.1).

Solute molecules get distributed in the hydrogen-bonded solvent, which resists the distribution. Such resistance causes opposing forces and gets a way out in the form of a track or virtual capillary model. It induces a shear stress generating friccohesity in transforming the CFs to intermolecular forces (IMFs) or distributing forces (Fig. 19.2).

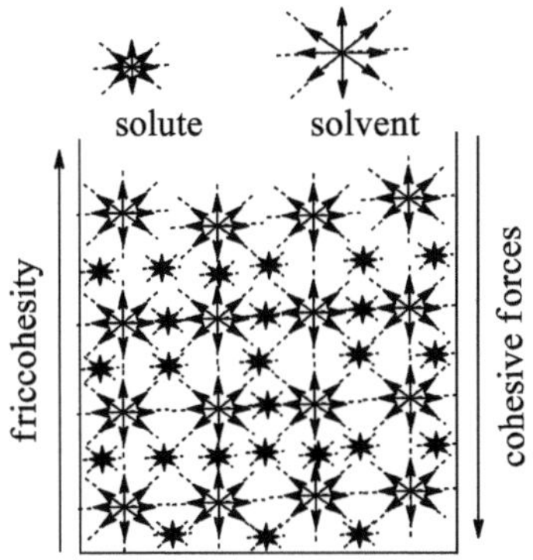

Figure 19.2 Relationship between friccohesity and cohesive forces of solvent vis-a-vis structural state of the resultant liquid mixture.

19.2 Two-Way to One-Way Chemical Activity: Multiple Kinetic Energy Indexes

Salts in aqueous solutions disrupt the water structure and align the monomer water molecules around the counterions. Such structural realignments allow water with opposite charges. The electrostatic poles of water molecules around the ions develop strong electrostatic interactions within the water dipoles. The ions remain at the center of the hydration sphere; they are called embedded ions and have a strong ionic field around them. Such arrangements strengthen CFs. Stronger CFs develop higher surface tension and viscosity both. A strong thin film at the surface is developed due to the strong hydrogen bonding (HB) between the oppositely charged hydration spheres around the oppositely charged ions. The oppositely charged ionic hydration sphere is bound with the higher potential energy with a stronger CF.

Compared to the science of hydration sphere the scenario is different in the case of surfactants that consist of the hydrophilic and hydrophobic parts. The hydrophilic part mildly disrupts the water structure, but the hydrophobic part is unable to disrupt the HB except for reorienting the structured water. Two chemical

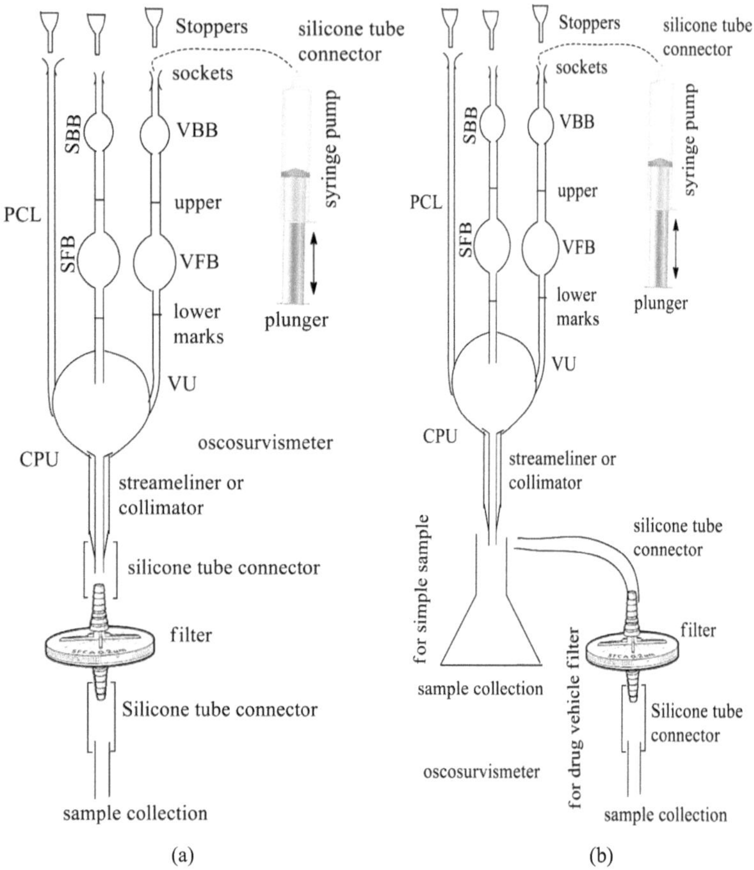

Figure 19.3 Structural frameworks of oscosurvismeter for measuring surface tension and visocsity by using filter membrane.

processes occur, where one involves hydrogen bond disruption with the hydrophilic head and the other involves the hydrophobic tail with non-hydrogen-bond disruption. The hydrophobic part reorients the structured water and aligns it around the alkyl chain. The reorientational process is entropic in nature, which helps the hydrophobic part to form a nanolayer around particles. The hydrophobic part causes Brownian motions, and several types of nanolayers induce different molecular motions. Such motions are rightly tracked by the heat capacity of the liquid mixtures at variable temperatures. In a nanoemulsion the surfactants surround the oil particles form globular structures. Each globular structure holds different energy levels, known as the multiple kinetic energy rule or index. So there is a shift from the two-way process to a one-way process if the dipolar solvent is added with salts and surfactant dispersion separately.

19.3 Compatibility in Light Wavelength and Particle Size: Theory

The size of a nanoparticle (NP) and the wavelength of light are mutually related. The size compatibility is proven as a sensor in the case of nanoemulsions like Rayleigh scattering. Under such scattering, the NPs smaller than ~250 nm in wavelength scatter laser light in all directions. A laser light source constitutes a monochromatic and coherent action that has a higher scattering intensity over time. So the small molecules undergo Brownian motions and the distance between the scattering particles constantly changes with time when constructive/destructive interference is developed by surrounding particles.

19.4 Gelatin Microemulsion as Viscosity Enhancer

Gelatin develops a thermodynamically stable, colorless or translucent microemulsion. The gelatin acts as a gelling agent in food, pharmaceuticals, photography, and cosmetic manufacturing, noted as gelatinous. It is derived from the collagen obtained from animal by-products. An irreversibly hydrolyzed form of collagen is used in gummy, gelatin desserts and some ice creams, dips, and yogurts. Gelatin sheets, granules, and powder are used in the food items, and they soak water from the food items. Gelatin is a mixture of the peptides and proteins produced by the partial hydrolysis of collagen extracted from animals. On hydrolysis, the natural molecular bonds between individual collagen strands break into a form that rearranges more easily and enhances viscosity and friccohesity both via a specific intermolecular multiple forces theory (IMMFT). The hydrolysis state of the gelatin is determined with the survismeter. The hydrolyzed state is altered on heating the gelatin at 25°C, 30°C, 35°C, 40°C, 45°C, 50°C, 55°C, and 60°C.

Question: Determine particle size and particle area using the survismeter.

Chapter 20

Hydrodynamic Size of a Gelatin Particle

20.1 Theory

Measure the viscosity η mPa·s of 0.1%, 0.5%, 1.0%, 1.5%, and 2.0% aqueous gelatin solutions prepared with Millipore water. The η of the sample is calculated as under:

$$\eta = \left[\left(\frac{t}{t_0}\right)\left(\frac{\rho}{\rho_0}\right)\right]\eta_0$$

Here, η_0 is viscosity of water, t and t_0 are viscous flow times, and ρ and ρ_0 densities of sample and water as solvent, respectively. However, friccohesity, which produces highly accurate results with the highest resolution, does not need density for its calculation. Relative viscosity η_r is calculated and processed as under:

$$\frac{\eta}{\eta_0} = \eta_\mathrm{r} = 1 + 2.5\varphi,\ \varphi = \frac{4}{3\pi r^3 N_\mathrm{A} c}$$

In the following equation, φ_i is the volume fraction of a constituent and V_i is volume divided by the volume of all the constituents of mixture, V, before mixing.

$$\varphi_i = \frac{V_i}{V}$$

The SI unit is m^3/m^3. The volume fraction is the volume concentration in an ideal solution. Because the volumes of the constituents

Survismeter: Fundamentals, Concepts, and Applications
Man Singh and Sunita Singh

ISBN 978-981-4774-70-3 (Hardcover), 978-0-429-02761-1 (eBook)
www.panstanford.com

are additive, the volume of the solution is equal to the sum of the volumes of its ingredients. The sum of all the volume fractions of a mixture is equal to 1.

$$\varphi_i = \sum_{i=1}^{n} V_i = V \quad \text{or} \quad \sum_{i=1}^{n} \varphi_i = 1$$

The solvent volume surrounding the gelatin particles is calculated and is fitted in the equation given as under to determine the hydrodynamic size of particles.

$$\varphi = \varphi^0 + \frac{4}{3\pi r^3 N_A c}, \quad \frac{3\varphi}{4\pi N_A c} = r^3, \quad r = \sqrt[3]{\frac{3\varphi}{4\pi N_A c}}$$

Here, φ is the volume fraction of the solvent surrounding the particle, c is the composition, N_A is the Avogadro number, and r is the hydrodynamic size.

20.2 Activation Energy

The energy that activates gelatin is derived from medium molecules and is activation energy that is determined by measuring the viscosity at 25°C, 30°C, and 35°C or the desired temperature.

$$\log \eta = \log A - \frac{E^*}{2.303RT}$$

Gibbs energy is calculated as under:

$$\log K_{eq} = -\frac{\Delta G}{RT} \quad \text{or} \quad \Delta G = -2.303RT \log K_{eq}$$

$$\text{or} \quad \Delta G = 2.303RT \log \eta$$

20.3 Unique Experiments: Theory and Method with the Survismeter

Determine the adsorption of CH_3COOH (acetic), CH_3CH_2COOH (propionic), $CH_3CH_2CH_2$ CH_2COOH (butyric), $CH_3CH_2CH_2CH_2COOH$ (pentanoic), and $CH_3CH_2CH_2CH_2CH_2COOH$ (hexanoic) acids on activated charcoal.

20.3.1 Theory and New Method

In general, the adsorption of adsorbates on adsorbents is determined by the titration method. The amount of adsorbate that should be adsorbed in the prescribed time span is ensured by titrating the filtered liquid because the liquid that was holding adsorbates like acetic acid could have been adsorbed by an adsorbent like activated charcoal. In the titration method a lot of chemicals are used, like phenolphthalein indicator and NaOH, and after neutralization of the solution the mixtures are discarded, which causes pollution. Therefore, a new method has been developed to determine the acetic acid adsorption by charcoal. It is seen that the surface excess concentration of the surfactant in water is determined by using the surface energy of a continuous thin film of the water at the air-water interface. The interface is developed by the hydrogen bonding (HB) of the water. Once the surfactant gets saturated at the interface then the surface energy becomes constant. Thus, a similar method could be extended to acetic acid adsorption by activated charcoal since activated charcoal has also got a saturation point for acetic acid adsorption that shows a constant surface energy. So the surface energy with time could be measured and when it comes to a constant point, saturation is attained. A series of acids is given in Fig. 20.1 for doing the adsorption experiments with.

Determine the size of the nanoemulsion particles in the nanoemulsions noted in Fig. 20.2. The hydrophobic–hydrophobic

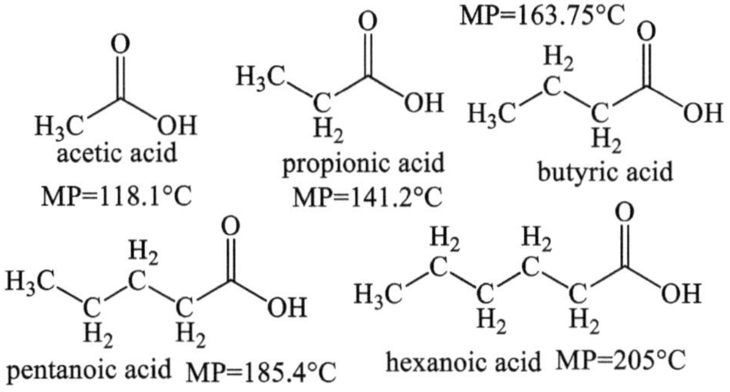

Figure 20.1 Structures of organic acids with increasing alkyl chain for study of hydrophobic and hydrophobic interactions.

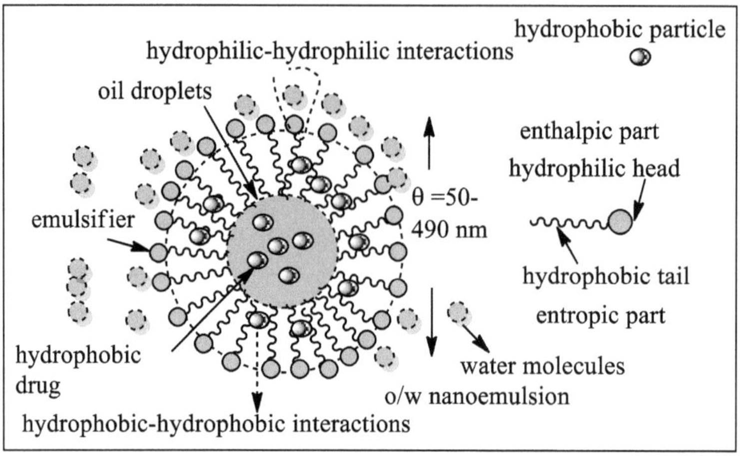

Figure 20.2 Spontaneity of hydrophobic molecules like oil or drug to move toward the hydrophobic core of normal micelles.

networking could also hold some structured water with the maximum entropic changes.

For propionic acids, the boiling point is 141.2°C, the molar mass is 74.07854 g/mol, the density is 990 kg/m^3, and the melting point (MP) is 21°C. For butyric acid the MP is 163.75°C. Therefore, the effect of the MP could be checked with the adsorbed amount by the activated charcoal. Pentanoic acid is a colorless liquid with a density 0.94 of g cm^{-3} and a boiling point of 185.4°C, and the MP of hexanoic acid is 205°C. Therefore, the nanoemulsion of melamine formaldehyde polyvinylpyrrolidone (MFP) resin could be developed in these solvents or volatile solvents for developing a nonporous thin film. The friccohesity data could define the thermodynamical stability of a nanoemulsion of the MFP resin for better efficiency of adhesive forces to be generated on the chosen surfaces.

Chapter 21

Surface Area of a Gelatin Particle

Surface tension (γ, mN/m) of the 0.1%, 0.5%, 1.0%, 1.5%, and 2.0% aqueous gelatin solutions is determined as under:

$$\gamma = \left[\left(\frac{n_0}{n}\right)\left(\frac{\rho}{\rho_0}\right)\right]\gamma_0$$

Here, γ_0 is the surface tension of water, n and n_0 are pdn, and ρ and ρ_0 are densities of sample and water as solvent, respectively. For surface excess concentration the data are fitted in the following equation:

$$-d\gamma = 2.303\ RT\tau \log c$$

The γ data are plotted against log c, which gives a slope value as follows:

$$\text{slope value} = 2.303\ RT\tau \quad \text{or} \quad \tau = \frac{\text{slope value}}{2.303\ RT}$$

The τ data are fitted in the following equation to determine the surface area of individual molecules:

$$\tau = \frac{\text{mol}}{\text{m}^2} \quad \text{or} \quad a = \frac{1}{\tau} \quad \text{or} \quad a = \frac{1}{\tau\frac{\text{mol}}{\text{m}^2}} \quad \text{or} \quad a = \frac{\text{m}^2}{\tau\text{mol}} \quad \text{or}$$

$$a = \frac{(1 \times 10^9\ \text{nm})^2}{\tau N_\text{A}} \quad \text{or} \quad a = \frac{1 \times 10^{18}}{\tau N_\text{A}}$$

Survismeter: Fundamentals, Concepts, and Applications
Man Singh and Sunita Singh

ISBN 978-981-4774-70-3 (Hardcover), 978-0-429-02761-1 (eBook)
www.panstanford.com

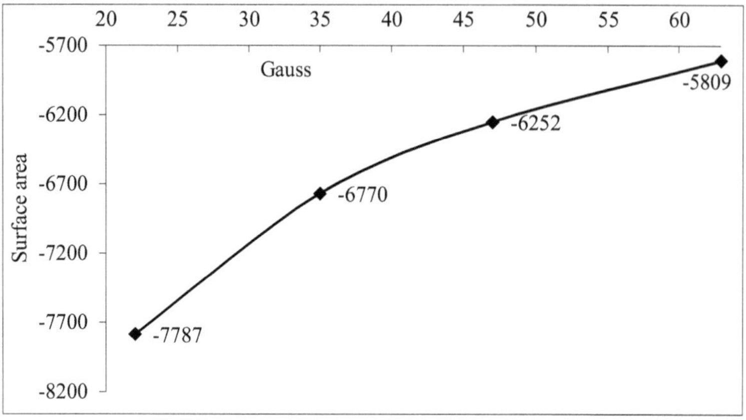

Figure 21.1 An increase in surface area of pepsin on exposing its aqueous solution to increasing magnetic field.

After a quantitative study of aqueous gelatin solutions the medium could be changed from an aqueous solution to aqueous ethanol, aqueous sorbitol, aqueous polyethylene glycol (PEG), aqueous tweens, and others as per industrial applications. The gelatin + dendrimers, gelatin + surfactants, gelatin + flavonoids, and gelatin + free radical in the aqueous medium could be studied. Such mixtures of specific friccohesity could act as the best extracting medium or dispersant. Gelatin of the photographic and pharmaceutical grade is obtained from beef bones and is dissolved in hot water and set to cool. It is restrictedly dissolved in cold water and is soluble in most polar solvents, like dimethyl sulfoxide (DMSO), dimethylformamide (DMF), and acetonitrile.

Figure 21.1 depicts an increase in the surface area ($1/\tau$, cm^2/mol) of 0.05 g% pepsin in aqueous solution. The solution was subjected to a magnetic field on giving 1.0, 1.8, 2.1, and 2.5 amp. (ampere) to cause 22, 35, 47, and 63 gauss with time interval (t, sec.), respectively [4]. Figure 21.1 depicts the unfolding of protein with a higher magnetic field due to polar peptide bonds that stretch and unfold the pepsin.

Therefore, apart from the magnetic field inducing a structural behavior of pepsin protein the gelatin solutions could induce viscoelastic flow and streaming birefringence.

21.1 Gelatin Fluid Dynamics

Viscosity and friccohesity measurements become attractive research work with gelatin solutions with fluorescent dyes and dendrimers. Gelatin mixtures could be studied with several alkali salts, surfactants, dendrimers, and lanthanides. Gelatin is dispersed in relatively concentrated acids and remains stable for ~15 days, with infinitesimal chemical changes. With higher thermodynamic and kinetic stability gelatin is useful for coating or extrusion into the precipitating bath provided there. Gelatin gels are sensitive to pH and temperature variations so the gels exist over a small temperature range from the melting point of the gel, depending on the gelatin grade and concentration (<35°C), which show a lower limit of freezing point with crystallization at ice point. The viscosity and friccohesity of gelatin–water mixtures are higher for higher gelatin concentrations at ~4°C. The gel strength is checked with the Bloom test (BT). The BT measures the strength of gelling or the gelatin. O. T. Bloom had developed and patented the test in 1925. The BT determines the weight (g) needed by a probe (diameter 0.5 in.) to deflect the surface of the gel by ~4 mm without breaking it. The result is expressed in Bloom grades, which fall in the range of 30 to 300 bloom. The BT is used for soft gels. For the BT a 6.67% gelatin solution is kept for 17–18 h at 10°C prior to being tested.

Similarly, the level of homogenization of gelatin with several solvents in the presence of surfactants could constitute a nanoemulsion for industrial applications for developing a heat and electricity conducting membrane. The affinities of gelatin nanoemulsions with several lanthanides vis-à-vis their protein docking could be a new research methodology. So a new research methodology for gelatin gel with additives like salts, bovine serum albumin (BSA), and dendrimers could constitute interesting solvodynamics. Therefore, for studying their physicochemical properties the one-step process used to determine friccohesity, intramolecular multiple forces theory (IMMFT), and tentropy with a survismeter could also be most effective for checking the quality of gelatin gels. For a viscous flow of aqueous gels, the magnitude F of this force is proportional to the speed u and the area A of each plate and inversely proportional to their separation y-axis.

$$F \propto A\frac{dv}{dy} \quad \text{or} \quad F = \eta A\frac{dv}{dy}$$

The proportionality factor η is dynamic viscosity, dv/dy is the rate of shear deformation or velocity noted as a derivative of fluid speed in a direction perpendicular to the plate. Newton had studied the expressions of viscous forces by the differential equation

$$\frac{F}{A} = \tau = \eta\frac{dv}{dy}.$$

Here, τ and dv/dy are the local shear velocity or velocity gradient and the fluid flows along the parallel lines and perpendicular to y-axis. The velocity may or may not vary linearly with the y-axis. The velocity does not vary for Newtonian liquids, but it varies for non-Newtonian liquids, which undergo certain transitional changes with variable orders with compositions. For the first-order liquid mixtures dv/dy remains unchanged because of the formation of regular liquid mixtures. However, if the liquid mixtures undergo changes during viscous flow, then dv/dy changes occur drastically. Therefore, the thermodynamic and kinetic stabilities could also be determined by the nature of the change in the viscosity. Also the unit surface area A of the inner surface of the capillary remains unchanged. Then a new equation is developed, as follows:

$$F \propto \eta \quad \text{or} \quad F = k\eta$$

The unit of viscosity is determined as under:

$$\frac{F}{A} = \tau = \eta\frac{dv}{dy} \quad \text{or} \quad \frac{\text{N}}{\text{m}^2} = \frac{\text{m}}{\text{s}}\cdot\frac{1}{\text{m}} \quad \text{or} \quad \eta = \text{N·s·m}^{-2}$$

Figure 21.2 explains the mechanism of viscosity obtained on viscous flow within a stationary and rigid capillary, noted as under. The mass transfer is from a higher potential energy to a lower potential energy within the uniform inner diameter (id) of the capillary and is referred to as viscosity. Similarly the iodine transfers from the

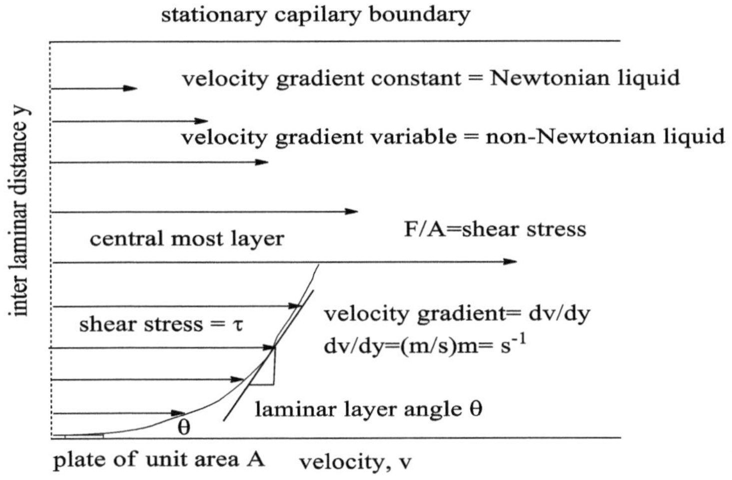

Figure 21.2 Trends of velocity gradients for fluid dynamics of liquid mixtures with time.

CCl_4 phase to the water phase within the liquid phases, which could be noted as the viscosity within liquid boundaries. It could be determined by calculating the transfer of iodine from the CCl_4 to the water phase within the known time, noted as under:

$$\text{iodine viscosity within liquid capillary} = \frac{\text{transferred iodine amount}}{\text{time taken, min.}}$$

In context of two immiscible solvents, the surface tension with the activities of a surface unit and also the viscosity with a unit area which induces frictional forces become significant. The interface of both the surface and the bulk phase constitute the most active physicochemical state. It develops a highly active surface which is noted as viscous or viscosity surface of a thin film which is formed on the solid surface of the capillary.

21.2 Friccohesity of $AgNO_3$ Colloidal Solution

Fluid dynamics of the colloidals furnish information about structural activities of the salt molecules, which develop colloidal solutions like $AgNO_3$, on mixing them in a warm gelatin solution containing KBr,

NaCl, or other alkali metal halides. The reaction precipitates into the fine light-sensitive insoluble silver halide crystals.

$$AgNO_3 + KBr = AgBr + KNO_3 \quad \text{and} \quad AgNO_3 + NaCl = AgCl + NaNO_3$$

AgX ($X = Cl^-$, Br^-, I^-halides) is peptized by the gelatin and a quantity of gelatin emulsion. The pH buffer, crystal habit modifier, metal dopants, ripener, ripening restrainer, surfactants, defoamer, emulsion stabilizer, and biocide are used in emulsion making. This approach develops immobilization in a one-step process. Most modern emulsions are washed to remove some of the reaction by-products (KNO_3 and excess salts). The washing step to desalt a mixture is attained by ultrafiltration, dialysis, coagulation (acylated gelatin), or a classic washing method. For making an emulsion, sensitivity is increased by using the chemical as sensitizing agents and sensitizing dyes. A device is noted in Fig. 21.3 that assists in superwashing of KNO_3 and excess salts. After every elution, friccohesity is determined. Constant friccohesity values indicate complete desalting.

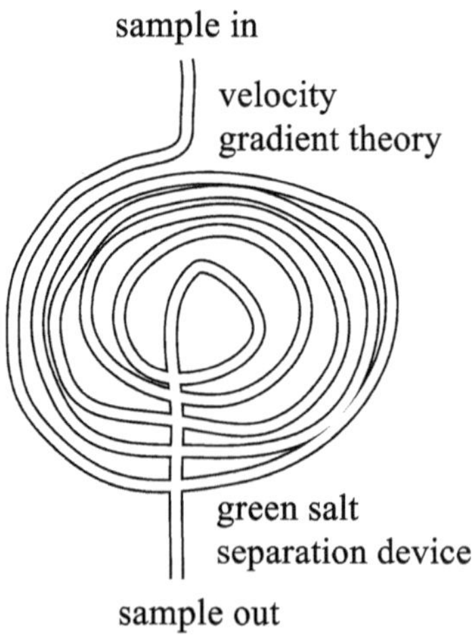

Figure 21.3 Inducing shear stress and strain on molecular species available in the liquid mixture by passing it through a capillary of uniform inner diameter. The friccohesity acts as a indicator as this arrangement is connected to survismeter.

21.3 Auto Purity Device

Water with added fiber: Fiber present in the water is not insoluble, but it remains active. If it is soluble then it may not sustain for longer.

Water for expectant mothers: Mixing extra calcium and folic acid into bottled water protects mother and baby both. The hydration is for the expectant mother but most pregnant women do not need as much fluid as nonpregnant women.

Question: Determine the packing of calcium and folic acid in bottled water using the survismeter.

21.4 Water for Athletes

This water contains sodium, potassium, amino acids, and some herbs, the amount depending on the sweating of the athlete. If an athlete sweats a lot for a long time extra sodium and potassium from the beverage is needed. Excess sodium is risky for a person with high blood pressure, and individuals with certain cardiac conditions or on certain medications have to avoid a high-potassium diet. Taurine, an amino acid, improves physical performance, and ginseng herb improves athletic agility.

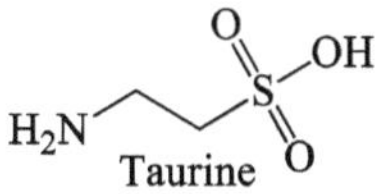

Taurine

Taurine, or 2-aminoethanesulfonic acid, as organic acid is found in animal tissues as a major constituent of bile. Taurine is ~0.1% of the total human body weight and plays a critical role in the conjugation of bile acids, antioxidation, osmoregulation, membrane stabilization, and modulation of calcium signaling. Taurine is essential for cardiovascular function and development and function of skeletal muscle, retina, and the central nervous system. Taurine contains H_2SO_4, while the majority of biologically occurring acids contain weaker acidic carboxyl groups. Fortified water contains sugar (which is similar to soft drinks, which have extra calories), along with vitamins or other nutrients, but compared to soft drinks, fortified water has less sugar and fewer calories. Low-calorie enhanced water with artificial sweeteners contains fiber and other phytochemicals and antioxidants for health safety. A conventional emulsion has droplets of 100 nm to 100 μm radii, which are thermodynamically unstable because of relatively large positive free energy (interfacial tension, IFT) associated with contact between the oil and aqueous phases. The emulsions remain optically opaque as the droplets have similar dimensions, like wavelength of light ($r \approx \lambda$), which scatter light strongly. The size induces a contrast in refractive index between the droplets and the surrounding liquid, which is not close to zero, and the droplet concentration is not too low.

Chapter 22

Optical and Molecular Interface

Nanoemulsion as a conventional emulsion contains droplets of 10 to 100 nm radii. A droplet relatively small in size as compared to the wavelength of light ($r \ll \lambda$) is transparent or slightly turbid (microemulsions). A small droplet size has better stability to gravitational separation and aggregation than emulsions with higher friccohesity. Such systems remain thermodynamically unstable since the separated oil and water phases have a lower binding energy than the emulsified oil and water phases. A microemulsion of 2 to 50 nm in size is thermodynamically stable with a lower free energy than that of phase-separated components of which it is comprised. These tend to form the microemulsion spontaneously. If a microemulsion under specific composition and temperature is altered then it is no longer thermodynamically stable because a microemulsion changes into a nanoemulsion, a conventional emulsion, a liquid crystalline system, or a phase-separated system. On bringing it back to the original conditions it should be a microemulsion, depending on kinetic energy barriers. Microemulsions are optically transparent as their particle size is much smaller than the wavelength of light ($r \ll \lambda$) and so light scattering is weak. Microemulsions are distinguished from nanoemulsions in terms of particle shape. Nanoemulsions form spherical droplets because the relatively high interfacial tension

Survismeter: Fundamentals, Concepts, and Applications
Man Singh and Sunita Singh

ISBN 978-981-4774-70-3 (Hardcover), 978-0-429-02761-1 (eBook)
www.panstanford.com

Table 22.1 Properties of emulsions prepared from oil, water, and an emulsifier

Emulsion	Radii	Stability	Surface to mass ratio/$m^2\ g^{-1}$	Appearance
Emulsion	100 nm–100 μm	Metastable	0.07–70	Turbid/opaque
Nanoemulsion	10–100 nm	Metastable	70–330	Clear/turbid
Microemulsion	2–50 nm	Stable	130–1300	Clear

(IFT) and small droplet size lead to a high Laplace pressure, favoring minimization of the oil–water interfacial area. Microemulsions contain spherical, ellipsoid, or worm-like shapes depending on the molecular geometry of surfactants used because the IFT is much lower. It distinguishes between them as it optimizes their formation, stability, physicochemical properties, and functional performance.

Nanoemulsions are thermodynamically unfavorable systems because of their positive Gibbs energy ($\Delta G > 0$), which is developed in creating oil–water interfaces. Naturally, nanoemulsions tend to break down over time due to gravitational separation, flocculation, coalescence, and Ostwald ripening. Nanoemulsions are comparatively more stable to gravitational separation, flocculation, and coalescence, but they are less stable to Ostwald ripening. For industrial applications nanoemulsions with higher kinetic stability are needed. Their kinetic stability is controlled by the composition of oil and water phases and microstructure based on particle size distribution or by incorporating stabilizers like emulsifiers, texture modifiers, weighting agents, or ripening retarders. The stabilizers used in the food industry to formulate nanoemulsions are brominated vegetable oil (BVO), long-chain triglycerides (LCT), whey protein isolate (WPI), and soyprotein isolate (soy). Emulsifier and surfactant adsorbed to oil–water interfaces as protective coatings around droplets prevent droplet aggregation like Tweens and Spans. The phospholipids are obtained from sources like egg, soy, or dairy lecithin, and amphiphilic proteins (WPI as whey protein isolate and caseinate) and polysaccharides like gum Arabic as modified starch are used to develop stable nanoemulsions. The texture modifier on increasing the viscosity or gel aqueous phase formation provides the textural characteristics or stops gravitational

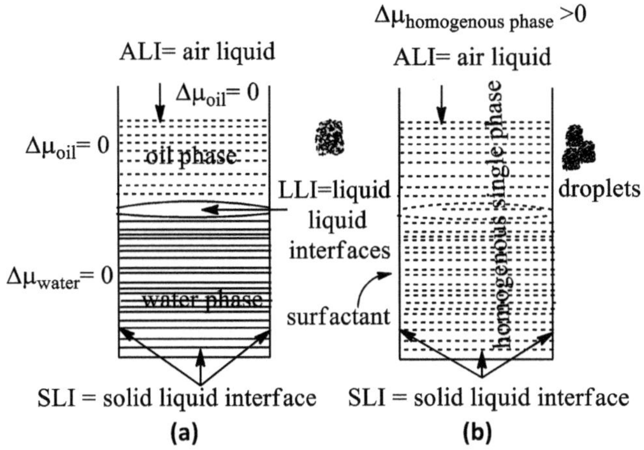

Figure 22.1 Oil–water homogeneous nanoemulsion formulation on increasing their mutual solubilities by adding suitable surfactants.

separation of sugars [sucrose and high-fructose corn syrup (HFCS)], polyols (glycerol and sorbitol), and polysaccharides (xanthan). Wetting agents are added to oil droplets to match their density with that of the surrounding continuous phase, like pectin, carrageenan, alginate, and proteins (gelatin, WPI, and SPI), and dense lipophilic materials [BVO, sucrose acetate isobutyrate (SAIB), and ester gums]. Ripening retarders as highly hydrophobic substances slow down Ostwald ripening when they are incorporated within oil droplets, like lipophilic materials with a very low water solubility (LCT and esters gums) (Figs. 22.1a,b).

22.1 Viscosity Increasing Theory

The selection of the most appropriate stabilizer to utilize within a nanoemulsion is useful in determining its fabrication and functional performance. Formation and stability of nanoemulsions depend highly on the nature of the oil phase used, like its polarity, water solubility, IFT, and rheology. A nanoemulsion's ability to contain small droplet sizes depends on the method used to create it: high energy or low energy. The methods used to prepare nanoemulsions, the physicochemical properties of nanoemulsions, and potential applications of nanoemulsions within the food industry are discussed. Water-soluble vitamins are passed into urine, so stronger binding of food materials with water is not suitable for digestion. Roughage noted as finer nanoemulsions is best to enhance bioavailability because these are released as there is no strong or weak binding among them. Excellent nanolets work to make items available for entering into the digestive system, where the body already has hydrophilic and hydrophobic (philicphobic) constituents to form nanoemulsion. The localized force stabilization poses a serious opposition to the flowing molecules over them so viscosity is created, which is interlinked with cohesive surface energy through friccohesity. To resolve and solve these effects to have a directional liquid potential for a fruitful prediction, the survismeter is an operating device. The oxygen gas adsorbing nanoemulsion or hydrocarbon holding the nanoemulsions could be explained by the friccohesity theory.

Chapter 23

Molecular Origin of Physicochemical Properties: IMMFT and Tentropy

The basic constituents of molecules are electrons, protons, and neutrons, where the neutrons do not participate in molecular activities listed in Fig. 23.1.

Figure 23.1 depicts how electrons share and develop molecules with unique patterns of electron arrangements or arrays in orbitals. Electrons are responsible for developing contacts within similar and dissimilar atoms. When electrons are equally shared then the molecules become most stable and do not cause much change in the structure on coming in contact with the solvent. In molecules the electronic arrangements are never equally distributed, which is because of electronegativity, size, electron density, or the nature of the atoms or groups, as listed in Fig. 23.2.

Molecules with confronting electronic states could act as free radicals in active mode, which constitute thermodynamics when they are dissolved in suitable solvents. Such electronically confronting molecules are noted as free radicals or peroxides containing unshared or nonbonding electrons or excess electron density in specific areas of molecules. This electronic state makes them extraordinarily active. Friccohesity originates from fluid dynamics

Survismeter: Fundamentals, Concepts, and Applications
Man Singh and Sunita Singh

ISBN 978-981-4774-70-3 (Hardcover), 978-0-429-02761-1 (eBook)
www.panstanford.com

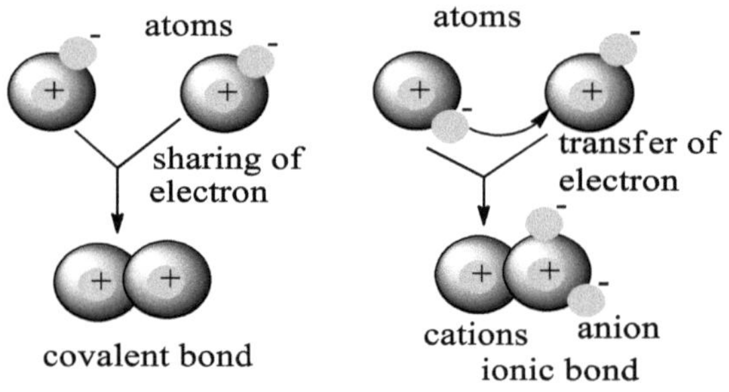

Figure 23.1 Potential energy of activities for sharing and nonsharing of electrons to form stable and nonstable bonds, respectively.

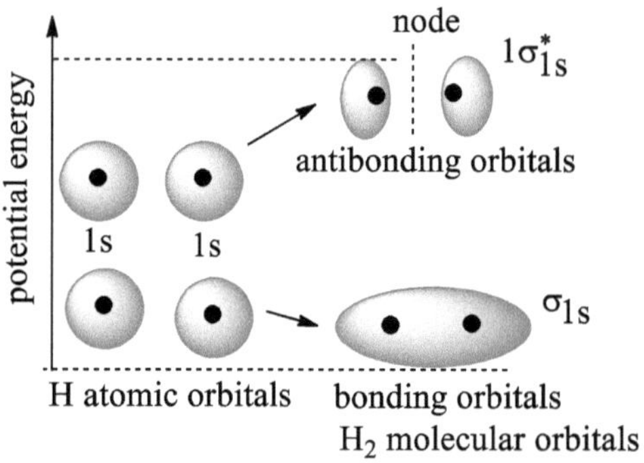

Figure 23.2 Potential energy of activities for sharing and nonsharing of electrons to form stable and nonstable bonds, respectively.

of such molecular mixtures in a suitable solvent, which decides the magnitude of friccohesity. Such electronic states favor a stronger distribution of solute and solvent mutually on producing higher viscosity and lowest possible friccohesity. Such features of electrons develop some charge differences on the atoms, which are noted in Fig. 23.2. Here, the antibonding elements are attached with molecule formation. The molecules on coming in contact with each other have additional energy to interact with the medium or molecules present in the vicinity of nearby molecules. But when molecules have equal sharing of electron density among their constituent atoms then the molecules become the most stable, with the highest potential energy and the least friccohesity (Fig. 23.3). For example, the N_2 molecule,

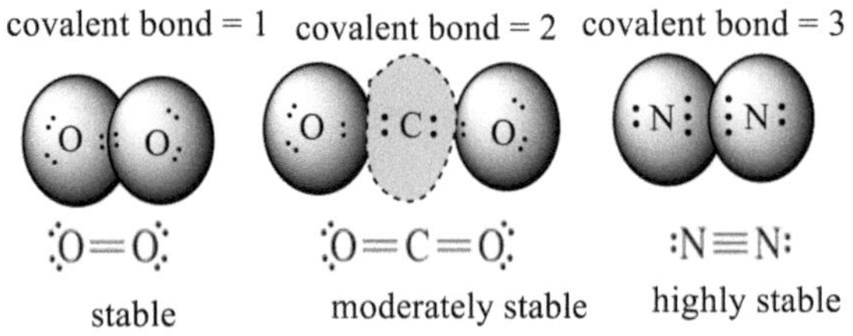

Figure 23.3 Stability of molecules on sharing of electrons with potential or binding energy.

with three covalent bonds, is the most stable and it does not interact even with UV rays despite the fact the atmosphere contains 78% N_2. It does not contribute to global warming.

Molecules during their formation phase occupy variable positions as per their electronic configuration, density, and unequal electron distribution or filling in the available orbitals as per Schrödinger equation within the boundary. Therefore, the above-mentioned science due to asymmetries in their structural frameworks leads to the establishment of the concept of intramolecular multiple forces theory (IMMFT) and tentropy. Under the IMMFT, the local canonical partitioning of the molecular spaces or the geometries having different potential energy sites induce locally different activities. Such activities induce entropy locally, where thousands of such localized ensembles inside the molecules remain operational. Therefore, the activities of the constituents of the molecules become important to visualize the valuable intramolecular activities that develop a relevance for friccohesity and its measurements with the survismeter. In such a process an electronic configuration of the atomic orbitals holding electrons in the molecules could be in the same phase or out of the phase that induces the physical state as well as the chemical activity of the resultant molecules (Fig. 23.4).

Atomic orbitals of such configurations seem to impart a mechanical strength to the molecules vis-à-vis their chemical bonds. Such movement of the molecules does cause some physical effects. For example, water vapor on moving causes impacts without affecting the H–O–H bond length or electronic configuration. The water molecules show an effective mechanical activity as well as their dynamics because when the molecules travel then they may

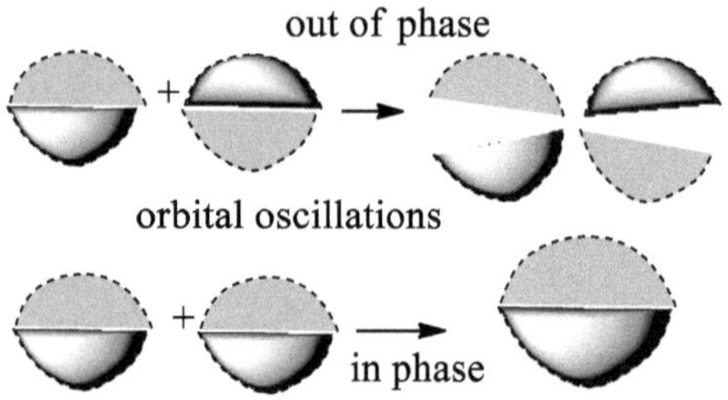

Figure 23.4 Intramolecular activities caused by orbital oscillations for electron induced molecular properties.

be in vapor form or in a liquid state that could induce motion-driven effects of atoms, like atomic motions with definite kinetic and thermodynamic potentials. Arrays, numbers of electronic configuration basically give birth to a concept of physicochemical sciences (unequal distribution of electronic charges, molecular masses, and motions) that could be termed as physicochemical chemistry or electron-induced molecular properties (EIMPs). The EIMPs are the basis for measuring the friccohesity, surface tension and viscosity, wetting coefficient, contact angle, and activation energy with the survismeter. The mechanical as well as dynamical impacts of molecules are monitored in the survismeter capillary.

Chapter 24

Thermodynamic Stability

Mechanical as well as dynamical impacts induce stability. For example, nanoemulsions are developed from oil and water interfaces with definite interfacial energy induced by surfactants. If similar interfacial energy is maintained over time then the nanoemulsions remain stable. The parameters that indicate a change remain the same, and no structural change in the oil particles is seen. If their friccohesity is measured at definite time intervals and it does not change in value with time, it depicts their thermodynamic stability along with structural stability. The major consideration for thermodynamic stability is that if the temperature decreases or increases in the infinitesimal range, no structural change occurs; this is similar in pattern to buffer systems, where the pH of systems remains the same on minor changes in H^+ and OH^- ions.

Accuracy and resolution: The most accurate data are produced by the survismeter. Several precautions and measures are taken in developing an unhindered sample flow. The focus is on components like structural groups, functional groups, electrostatic force, electronegativity, and molecular motions.

Survismeter: Fundamentals, Concepts, and Applications
Man Singh and Sunita Singh

ISBN 978-981-4774-70-3 (Hardcover), 978-0-429-02761-1 (eBook)
www.panstanford.com

24.1 Friccohesity Law of Interacting Surface Area

Friccohesity establishes an equilibrium between an interacting surface area of molecules holding interacting coordinates and an available surface area of the solid capillary of the survismeter for the solid–liquid interface (SLI). Such law is called the friccohesity law of interacting surface area.

Surface tension: The surface area and quality of solid glass surface in terms of the vacant forces and wetting ability of the surface matter a lot in developing accuracy in data. When the liquid sample wets out a solid glass surface then work is done on consuming the cohesive surface energy of the liquid sample and van der Waals forces are utilized. It could develop novel interacting interfaces of graphene–curcumin molecular dynamics of biomedical interest and proteins, DNA, and bacterial growth are studied using friccohesity.

24.2 Thin Film and Wetting of Solid Glass Surface

Stability in thin film forming and wetting activities is a fundamental science which develop an accuracy in the surface tension and also reproducibility in surface tension data. Figure 24.1 depicts the adhesive forces that work to form pendant drops at the cost of cohesive surface energy of the liquid.

So a high-quality glass capillary with a uniform surface area and forces is useful. The survismeter science works on the basis of SLIs as the liquid is permitted to move up and come down in the same capillary as a reversible process.

24.3 Study of Ion-Induced Dipole Forces

An ion-induced dipole attraction is a weak attraction that results when the approach of an ion induces a dipole in an atom or in a nonpolar molecule by disturbing the arrangement of the electrons in the nonpolar species (Fig. 24.2).

The molecules when brought in contact with other molecules induce molecular motions as noted in Fig. 24.3.

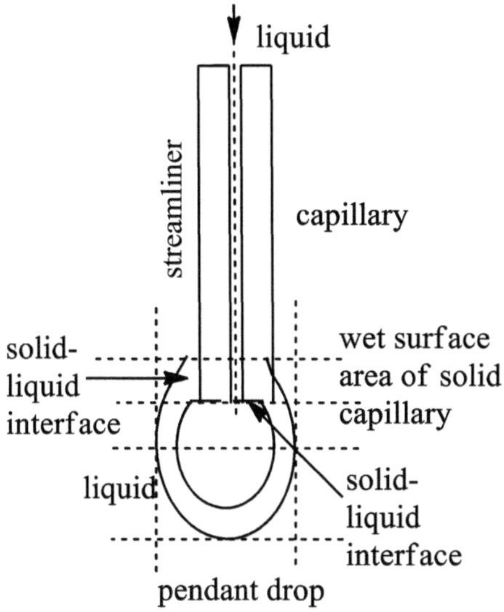

Figure 24.1 Basic physicochemical properties of liquid sample and dimensional geometry of solid capillary for forming pendant drops.

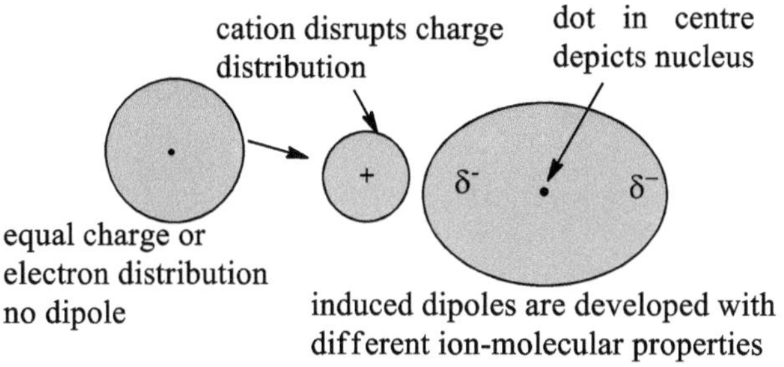

Figure 24.2 Mechanism of induced dipoles which affect shear stress and strain during fluid dynamics.

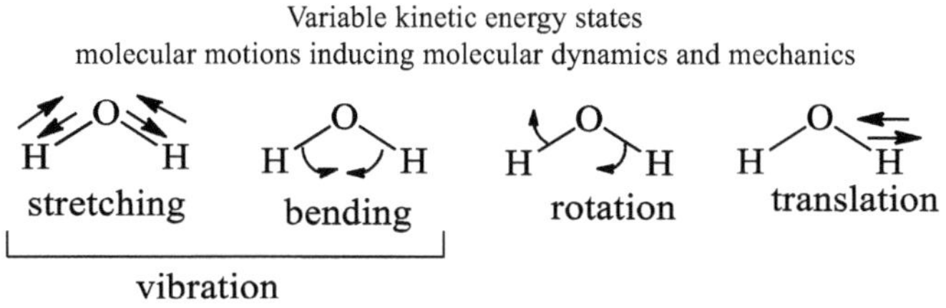

Figure 24.3 Atomic oscillations of water molecules causing variable molecular motions.

Chapter 25

Molecular Mechanics

The molecular motions listed in Fig. 25.1 constitute molecular mechanics based on classical mechanics of model molecular systems. The potential energy of working systems in molecular mechanics is calculated using the force fields theory. The molecular mechanics is useful for dealing with small molecules and large biological systems or material assemblies with thousands of millions of atoms. The atomic molecular mechanics are listed as under:

- Each atom acts as a simulated single particle.
- Each particle is assigned a radius (van der Waals radius), polarizability, and a constant net charge (quantum calculation or experiment).
- Bonded interactions act as springs, with an equilibrium distance equal to experimental or calculated bond length.
- Their spatial motions affect the tentropic activities.

So several interacting variations on this theme are explored, like many simulations have historically used an integrated atom where each terminal methyl group or intermediate methylene unit is considered a single particle. The large protein systems are simulated using a bead model that assigns two to four particles per amino acid. The electron cloud could be noted as under as per its electron density.

Survismeter: Fundamentals, Concepts, and Applications
Man Singh and Sunita Singh

ISBN 978-981-4774-70-3 (Hardcover), 978-0-429-02761-1 (eBook)
www.panstanford.com

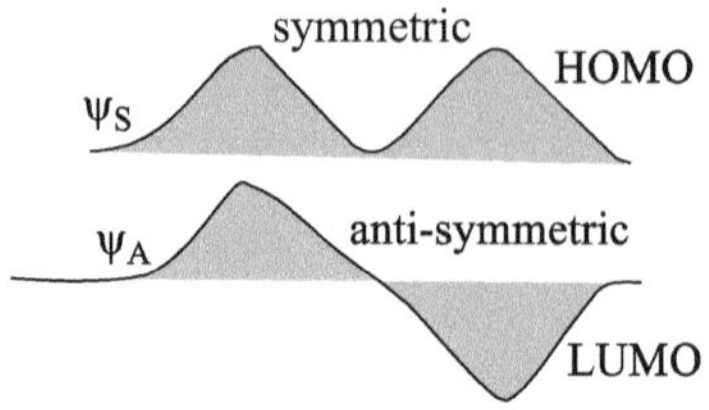

Figure 25.1 Representations of symmetric (HOMO = highest occupied molecular orbital) and antisymmetric (LUMO = lowest unoccupied molecular orbital) molecular orbitals with symmetric (ψ_S) or bonding and asymmetric (ψ_A) or antibonding wave functions, respectively.

This model is used to develop the grounds for molecular mechanics and dynamic entities with symmetric or highest occupied molecular orbitals (HOMOs) and lowest unoccupied molecular orbitals (LUMOs). Figure 25.1 explains hydrophobicity and hydrophilicity as per the size of the alkyl chain and the hydrophilic head. Methyltrioctylammonium chloride (MTOAC) is the most hydrophobic surfactant with the lowest surface tension and a high viscosity (Fig. 25.2) producing high friccohesity values. Such interacting activities of MTOAC signify the role of the larger population of sigma bonds [8 × 3 (octyl chain) = 24σ bonds] in its alkyl chain (Fig. 25.3). Probably, it proves that the greater hydrophobicity in a trifurcated form of the alkyl chain induces steric effects. Such disrupted numbers of 24 sigma bonds over a single alkyl chain find develop a vacancy for generation of higher friccohesity with more wetting abilities (Fig. 25.2).

The liquid–liquid interfaces (LLIs), orbital interfaces, and structural interfaces based on the hydrophobicity and hydrophilicity

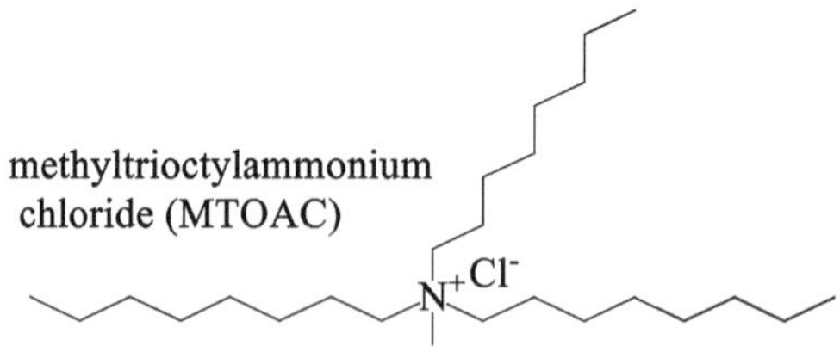

Figure 25.2 Hydrophobicity and hydrophilicity ratios in terms of three alkyl chains and chloride anion attached to quaternary nitrogen.

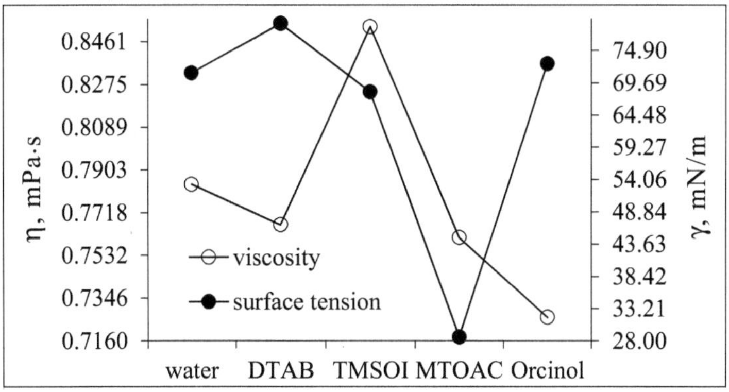

Figure 25.3 Opposite trends of viscosities and surface tension data of cationic surfactants in their aqueous solutions.

of the solvents could induce a docking-like activity of them. In such processes the distribution of the molecular forces is aligned around the atoms through chemical bonds and is critical and a deciding factor for the molecular activities. Thereby the electronic distribution could induce a structural hybridization within the molecular frameworks.

25.1 Molecular Dynamics

Molecular dynamics (MD) simulates the movements of *n* number of atoms in molecules, which are noted as the *n*-body simulation systems. Both atoms and molecules interact for a certain period of time to optimize the atomic motions with a definite spin and tentropy. Such atomic systems develop trajectories of the interacting particles (atoms and molecules), which are determined by the equation of the motion. The working forces between the particles and their potential energy are determined by the interatomic or molecular mechanics, which create their own force fields. Such molecular simulations are substantially applied in chemical physics, materials science, and modeling of biomolecules. Therefore, it is impossible to analyze the properties of the molecular complex systems consisting of a vast number of particles or the atom.

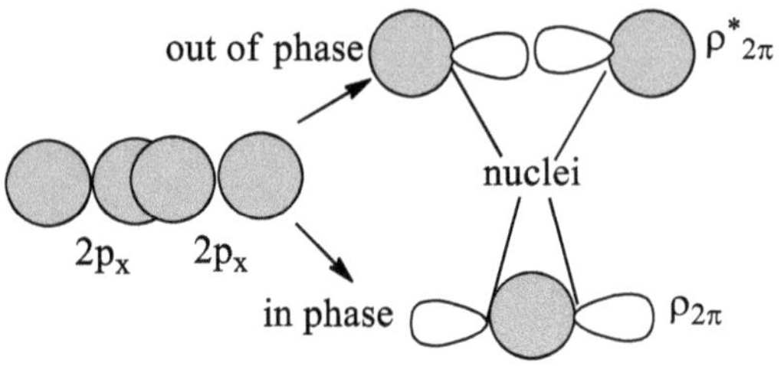

Figure 25.4 Sharing of electronic orbitals for compound formation on in phase and out of phase sharing.

Performing the MD simulation by using the numerical methods solves such complications. Longer MD simulations have certain restrictions and develop cumulative errors in the numerical integration, which are minimized with suitable algorithms and parameters. The ergodic hypothesis, which is applied for solving a single MD simulation, determines the macroscopic thermodynamic properties on integrating the possible canonical ensembles. The time averages of an ergodic system correspond to the microcanonical ensemble averages. The MD is a statistical mechanics by adjoining numbers of electrons and Laplace's vision of Newtonian mechanics to predict the animating forces and allow an insight into the molecular motions on an atomic scale. Figures 25.4–25.6 explain how atomic interactions can get optimized as per their electronic spins, which may be out of phase or in phase. Such spin activities lead to the development of the energy states involved in bonding and conceptualize the wave (ψ, chi) model of atomic orbitals.

The wave model of atomic orbitals holding electrons with electrons density (Figs. 25.5a,b) infer stability of the resultant molecules or robustness of a electronically chemical process.

Generalizing the concept of the chi effect, which is noted in Fig. 25.6, causes wave-type charge arrangements or ionic potential and spatial wave functions (ψ).

The split of energy as depicted in Fig. 25.6 signifies the concepts of tentropy and intramolecular multiple forces theory (IMMFT) both for homogenizing the energy distribution and the electron charge density, which get equilibrated throughout the atomic orbital and an optimized structure results. These orbitals develop molecules

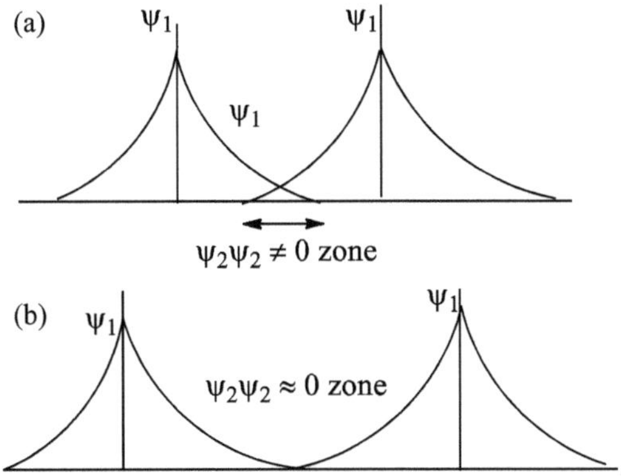

Figure 25.5 (a) The structure infers weaker overlapping of electronic orbitals. (b) The structure depicts maximum hybridization or overlapping of atomic orbitals with higher potential energy.

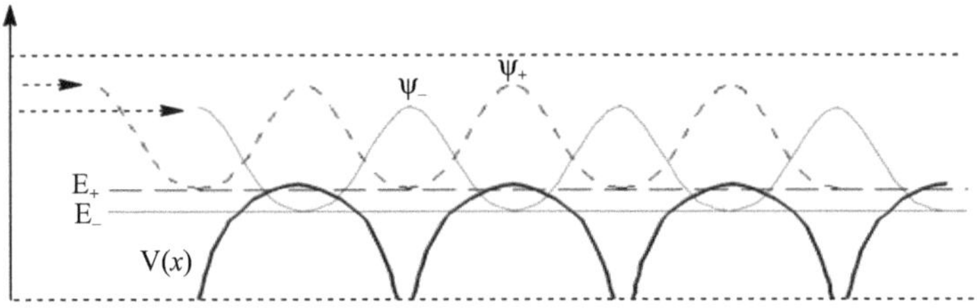

Figure 25.6 Ionic potential, spatial wave functions ψ split in energy.

by adjoining through chemical bonds, which induce a possibility for influencing molecular motions. Such structural ideas formulate a substantial basis of the tentropy as well as of IMMFT concept materialized through the structural frameworks. The above-said concepts are experimented by subjecting the molecules to fluid dynamics which express specific data of friccohesity. Thus the model of friccohesity and cohesive forces (CFs) is explained through the following structural formulations working as substantial guidelines:

- New scientific insights as the new philosophy and facility
- New approach, new model, and new state of art
- New inspiration steps for developing a new science

- Experiments for microfluidics and friccohesity with survismeter
- Initial working idea
- Science and creative models of nanoemulsion formulation
- Conceptualization of the survismeter idea and arrays of survismeter science
- An asset for solution chemistry and nanochemistry chemistry
- Addition to science and technology and science single surface of coin
- Creative genesis of survismeter
- Conceptualization of the survismeter idea and invention as an asset for upper critical solution temperature (UCST)

Chapter 26

Domain of Physicochemical Processes

The emergence of survismeter science builds the grounds for creativity in the case of molecular combinations, which are materialized through chemical processes, where the structure of molecules like stereochemistry is tracked with the survismeter. For example, ionic liquids and dendrimers both induce specific interacting patterns. The patterns of the saturated, unsaturated, stereoisomer, and polymers have specific cohesive surface energy and frictional forces (FFs). In the chemical process the procurement of transport of energy occurs either from UV radiation or through bond breaking. However, in physicochemical processes, the chemical bonds are not broken in a single step since the energy of activation is gained in the following steps: bond stretching, initiation of disruptions through atomic motions, electronic motions, and electron proton interactions. So the study of the physicochemical properties is of use.

***Spectroscopy*:** External sources of energy absorption induce motions like electronic, rotational, vibrational, and reorientational to check the capability of molecules within the specific condition.

Survismeter: Fundamentals, Concepts, and Applications
Man Singh and Sunita Singh

ISBN 978-981-4774-70-3 (Hardcover), 978-0-429-02761-1 (eBook)
www.panstanford.com

26.1 Fundamentals of Physicochemical States

Mixing of two separate molecular entities together does not develop liquid mixtures. If the electron clouds of each bonding atoms are strongly integrated with sigma bonds that have atoms of similar electronegativity, then a mixture is not developed and both molecules remain insoluble. No concept of physicochemical properties could have been generated with such molecules without energy exchange within molecular structures. Thereby, the solvent acts as a source for activation energy, which induces and develops effective molecular dynamics. Similar effects are induced in molecular bonds as they respond to UV radiation, infrared (IR) radiation, flame ionization detector (FID), and thermionic detector (TID). So a common factor is the induction of molecular response in terms of motion.

IR Sensor

A transmission medium is required for IR transmission that consists of vacuum or an atmosphere or an optical fiber. Optical lenses made up of quartz, CaF_2, Ge and Si, polyethylene Fresnel lenses, and Al or Au mirrors converge or focus IR radiation. To control the spectral rays band-pass filters are applied. Fourier-transform infrared (FTIR) detectors detect a radiation that has been in focus. The output from a detector is very small so preamplifiers coupled with electric circuit are required to further process the received signals. IR tracking or IR homing is a missile guidance system that operates using IR electromagnetic radiation emitted from a target for tracking it. Such missile systems act as heat seekers as IR radiation is radiated strongly by hot bodies like vehicles and aircraft.

26.2 Art History and Restoration

IR imaging of artwork for analysis and restoration is known as IR reflectography and is used by art historians to reveal hidden layers in paintings. Such molecular distribution, noted in Figs. 26.1(a–d), affects the magnitude of friccohesity. The molecules of variable structure and geometry distribute and develop variable canonical

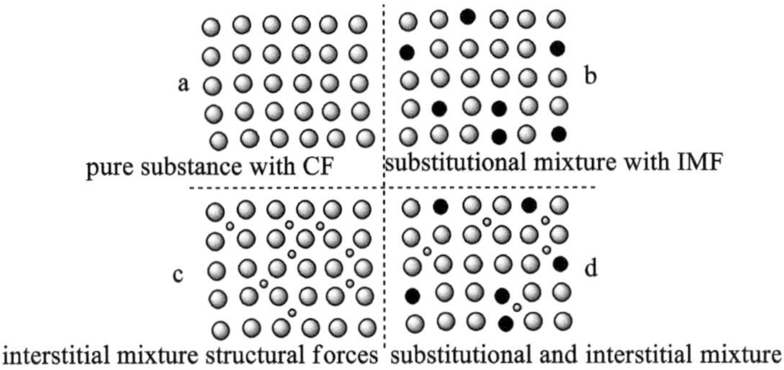

Figure 26.1 The patterns of molecular distributions with several molecular models like (a) CF (cohesive forces), (b) IMF (intermolecular forces), (c) interstitial structural forces, and (d) substitutional and interstitial mechanisms for nanoemulsion formulations.

ensembles, inducing specific tentropy and intramolecular multiple forces theory (IMMFT).

***Gel thermometer*:** Polymethyl methacrylate (PMMA), starch, and polyethylene glycol (PEG) gel when filled in a capillary show a volume change when the temperature varies.

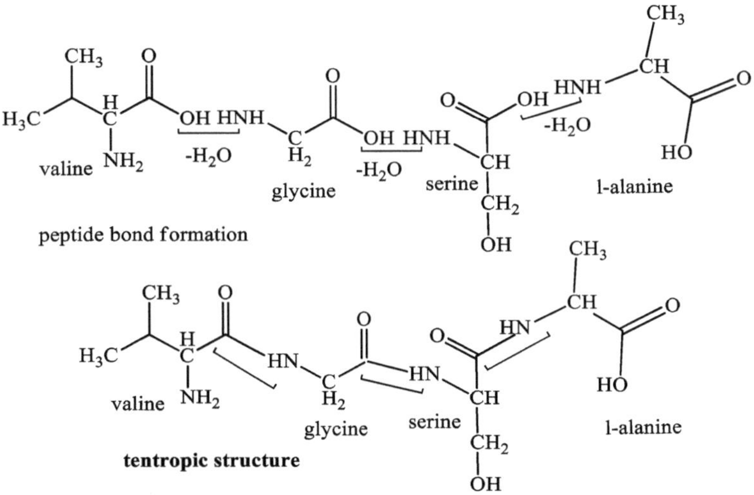

Figure 26.2 Molecular constituents and localized partitioning with uncontrolled degenerated orders which systematize intramolecularly, causing tentropy expressed by friccohesity.

Valine-glycine-serine-alanine tetrapeptide combines with the amino end of L-valine and the carboxyl end groups of L-alanine to form peptides (Fig. 26.2). It is a significant example of the actual meaning of the friccohesity concept where the cohesive surface energy or forces or cohesive forces (CFs) and the bulk binding forces (BBF) are responsible for the FFs where the localized surface forces become functional.

Chapter 27

Molecular Swirling

The survismeter, a unique scientific device, measures tension, viscosity, interfacial tension (IFT), wetting coefficient, and physicochemical properties simultaneously with high resolution and reproducibility. It saves laboratory resources by up to 95% through sequential multiuse of the liquid sample in its experimental methodology. The survismeter has the provision of a separate probe for surface tension, IFT, viscosity, and wetting coefficient, calibrated with standards of the National Physical Laboratory, India. The surface tension of a volatile sample at variable temperature is measured as the sample is jacketed and permitted to develop pendant drops in the controlled pressure unit (CPU) or the selection bulb of the survismeter. The selection bulb guides the liquid sample into the sample analysis probes.

27.1 Swirling Sound Generation

Swirling is initiated in processes of fermentation, solubilizing, and dispersion energy creation. A swirling sound comes on dissolving the specific molecules in the specific solvent. So swirling sound generation (SSG) could identify NaCl or KCl, glucose, and others. The

Survismeter: Fundamentals, Concepts, and Applications
Man Singh and Sunita Singh

ISBN 978-981-4774-70-3 (Hardcover), 978-0-429-02761-1 (eBook)
www.panstanford.com

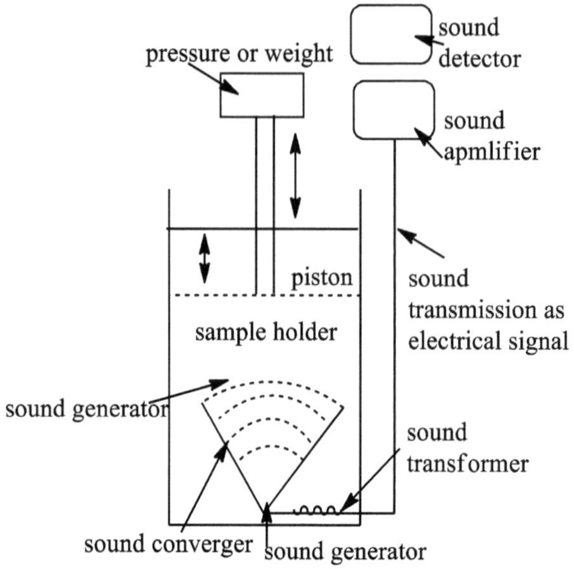

Figure 27.1 Identification of chemical substances through their swirling sound.

mechanism is illustrated in Fig. 27.1 and the equipment could be set up for measurements as illustrated in the figure. All parts and detectors are put sequentially for better working.

Naturally each molecule has its specific science, which causes a fixed force area or an activity domain acts as an active surface. Thereby, the intrinsic constitutional make up of molecules induce mechanical, electrical, and electronic forces that allow the moving fluid layer or sheet to slide over an adjacent layer. Thus it faces opposing forces. The measurements of the opposing forces cause a frictional movement. The advanced version with transducers and other accessories is noted in Fig. 27.2.

27.2 Physicochemical Properties of a Transitional Structure

Transitional structures that align and develop a new set of atomic arrangements with the new coordinates could configure with

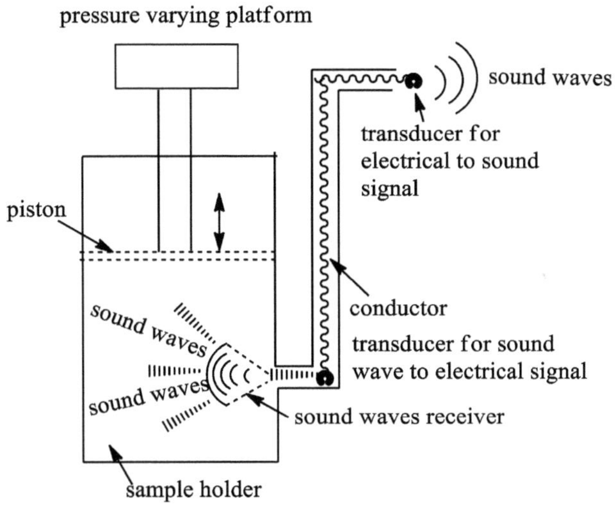

Figure 27.2 Identification of chemical substances by visually challenged.

electronic localization. The bulk as well as the surface forces undergo changes, for example, in a reaction where the reactants undergo structural changes to develop the new faces of the atoms in terms of the new molecules. The atomic and electronic bonds are involved in both cases. A hypothesis that induces correct thinking, dreaming, and successful invention gains importance for conducting useful experiments. Thus the swirling sound generator is useful, especially for visually challenged people.

Chapter 28

Fundamentals of Nanoemulsion

- Determine and identify the hydrophobicity and hydrophilicity of the Tweens 20, 40, and 60 in aqueous liquid mixtures from friccohesity values.
- Determine ion hydration (IH) and molecular hydration (MH) on electrostatic charge alignments and compare the values of both IH and MH.
- Determine cohesive surface energy of sodium dodecyl sulphate (SDS), dodecyltrimethylammonium bromide (DTAB), and methyltrioctylammonium chloride (MTOAC).
- Develop a standard calibration curve for philic-phobic versus surfactants to be used as a standard to determine the hydrophobicity and hydrophilicity ratio.
- Determine ionic size from friccohesity data of monocationic salts or molecules like surfactants measured with the survismeter.
- Determine ionic size from friccohesity data of divalent cationic salts measured with the survismeter salts.

28.1 Immiscible Solvent Phases

Immiscible solvent phases like oil and water develop liquid–liquid interfaces (LLIs) as a unique model to determine the

Survismeter: Fundamentals, Concepts, and Applications
Man Singh and Sunita Singh

ISBN 978-981-4774-70-3 (Hardcover), 978-0-429-02761-1 (eBook)
www.panstanford.com

interfacial tension (IFT) with variable chemical potentials of phases. The surfactants induce mutual miscibility on causing hydrophilic and hydrophobic interactions. Surfactants decrease the chemical potential of the individual solvent due to interactions as per liquid–liquid interfacial model solvents. The mutual miscibility of the solvents develops mixed solvents of a controlled polarity, density, and refractive index for biomedical and biophysical uses. The IFT at the LLI makes an energetic zone and catalyzes the growth of certain enzymatic processes using biomolecules due to the availability of energy at the interface. The study of the viscosities of liquids on mutual miscibility is a challenging task. Such liquids could be Newtonian or non-Newtonian and have a basic difference in kinetic energy. Emulsions have a higher kinetic energy as compared to solutions due to Brownian motions, which cause self-deriving forces with a gradient in chemical potentials. The electronic configuration of surfactants with functional groups and alkyl chains generates chemical potentials due to inductive and steric effects. Electronegativity of the O^-, I^-, and S^- of trimethylsulphoxonium iodide (TMSOI); the Cl^- and N^+ of methyltrioctylammonium chloride (MTOAC); the N^+ and Br^- of dodecyltrimethylammonium bromide (DTAB); and OH^- groups and π-conjugation of 3,5-dihydroxytoluene (orcinol) determine mutual wetting of benzene and water both. These models are thermodynamically active along with the electronegativities of their functional units. A methyl group with surfactants causes hydrophobic interactions, and positively charged N^+ or S^+ atoms develop hydrophilic interactions. So the mutual mixing is controlled by a ratio of hydrophobic and hydrophilic interactions. Surface tension, viscosity, and IFT depict the integrated behavior of the interactions. A study on the effect of buoyancy of a medium is useful, but limited studies have been conducted. Both MTOAC and TMSOI decrease the IFT of benzene + water interfaces by 42.67 and 3.6 mN/m respectively. Their effects on cloud points of water and phenol phases have been studied and reported elsewhere [5]. Viscosity, surface tension, and IFT data are of industrial significance in areas of cosmetics, extraction, and pharmaceuticals. Ginger, garlic, beet, spinach, fenugreek, coriander, black salt, and thymol seeds form organic microemulsions.

28.2 Anticaking Chemicals

Anticaking chemicals prevent lump formation or caking, allowing packaging, transport, and consumption of foods. Sodium aluminosilicate as an anticaking agent is added to table salts and dried milk, egg mixes, sugar products, and flours. In some countries sodium ferrocyanide and potassium ferrocyanide are the usual anticaking agents in table salt. Calcium and magnesium carbonates as natural anticaking chemicals are used in expensive table salts. There are water-soluble and alcohol-soluble anticaking agents, and they function either by absorbing excess moisture or by coating particles and making them water repellent. Calcium silicate ($CaSiO_3$), a common anticaking agent, if added to salt absorbs both water and oil. So an experiment is conducted to prioritize the absorption of water and oil by $CaSiO_3$ in a two-phase system by hanging $CaSiO_3$ at the boundary of water and oil phases.

28.3 Experiment for Simultaneous Adsorption of Water and Oil at the LLI

Procedure: Take 1 g of $CaSiO_3$, pack it in a teabag, and hang it vertically in water and oil phases separately (Fig. 28.1). After every

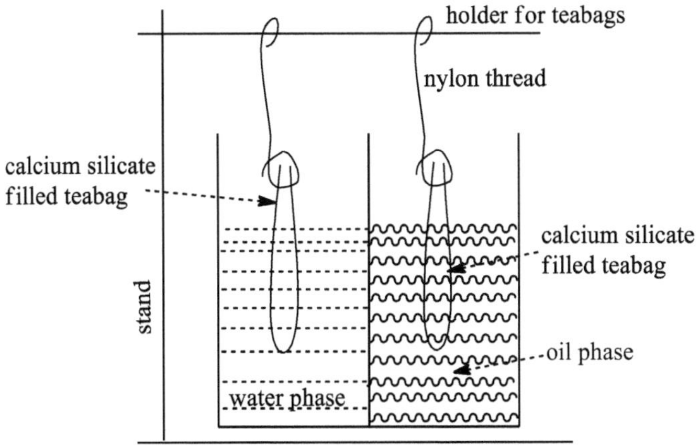

Figure 28.1 Two phase system with calcium silicate powder at their LLI.

15 min. withdraw the teabag and dry and weigh it. Measure the difference in weights. Make a comparison of the water phase weight and the oil phase weight. Such changes could be further determined by adding certain surfactants. Friccohesity on absorption could be studied by determining the friccohesity with the survismeter.

Chapter 29

Natural Preservative and Antioxidants

Food acids like vinegar, lemon juice (citric acid), tartaric acid, malic acid, fumaric acid, and lactic acid behave as stable food acids.

***Antifoaming agents*:** They develop higher viscosity and lower surface tension with higher friccohesity on reducing the foaming in foods. Similarly, antioxidants like curcumin and vitamin C act as preservatives by inhibiting the effects of oxygen on food and are beneficial to health. A curcumin nanoemulsion prevents oxidation as there is no place to allow O_2 gas in the mixture. Metallic nanoparticles (NPs) could also be protected from redox activities by capping them with a suitable agent.

***Bulking agents*:** Starch increases the bulk of a food item without affecting its nutritional value.

***Emulsifiers*:** Emulsifiers mix water and oils together to form nanoemulsions, like in mayonnaise, ice cream, and homogenized milk.

***Flavors*:** Flavors as food additives develop a particular taste or smell and may be derived from natural ingredients or created artificially. Humectants prevent foods from drying out.

Survismeter: Fundamentals, Concepts, and Applications
Man Singh and Sunita Singh

ISBN 978-981-4774-70-3 (Hardcover), 978-0-429-02761-1 (eBook)
www.panstanford.com

29.1 Organic Preservatives

These are extracts of lemon juice, fenugreek, ginger juice, garlic juice, chili juice, and curcumin. These inhibit spoilage of food due to fungi, bacteria, and other microorganisms. Thickeners increase viscosity without changing structure.

29.2 Photon Conversion into Sound of a Specific Frequency

Photographic film (PGF) traps photons and converts them into sound of a specific frequency. It identifies the chemical molecules attached on the PGF. The photons on absorption by the AgX crystal push electrons to go to a conduction band (a delocalized electron orbital with a higher energy than the valence band). The electrons are attracted by a sensitivity speck as a shallow electron trap. It could be a crystalline defect or a cluster of AgS, Au, and other trace elements (dopant) combined with an interstitial silver ion to form a silver metal speck.

29.3 Chemical Activities During Liquid Flow in the Survismeter Capillary

Several chemical activities and geometrical positions could be attained on viscous flow for measuring surface tension, interfacial tension (IFT), cohesive surface energy, stretching tension at the air–liquid surface, wetting coefficient, contact angle, surface energy, viscosity, solvent entanglement, polymer molecular weight, activation energy, solvent binding, particle size, rheology, food texture, sedimentation coefficient, coalescence, flocculation, fine impurity aggregation, dispersion, pesticides emulsions, and inkjets. The geometrical activities of the atoms in the molecules could affect the interacting molecular interfaces as shown in Figs. 29.1(a–c).

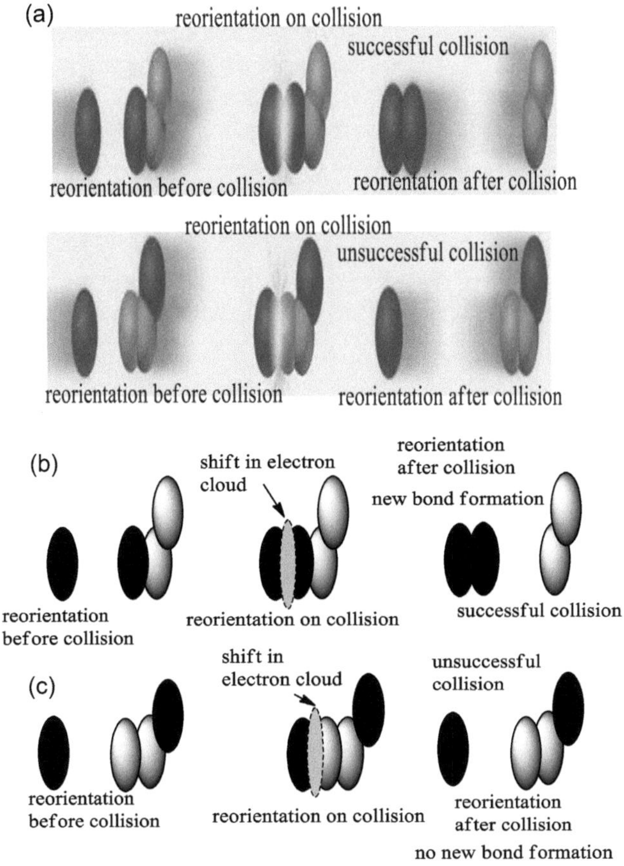

Figure 29.1 Reorientational, electronic configuration, collisions of molecules vis-a-vis medium molecules develop a pattern of fluid dynamics.

Such interfaces of molecular geometries could be helpful in understanding the chemistry of the drug–dendrimer interfaces (CDDI) if it is explored by using the oscosurvismeter.

Chapter 30

Cohesive Forces, Friccohesity, and Crop Growth

Stronger cohesive forces (CFs) inhibit water availability to roots. This mechanism of the solvent, of binding to enzymes or receptors, is applicable in pharmaceutical drugs. It increases or decreases the ability of the enzyme to function. Solvent or dendrimer binding to a protein is steric or spatial, involving various noncovalent interactions, contrary to covalent drugs binding on changing an active site (irreversible inhibitors). The lock-and-key model of enzyme binding is an efficient approach so long as the drug (key) is the proper dimension to fit the enzyme binding site (lock). On an appropriate-sized molecular scaffold, drugs interact with the enzyme noncovalently in order to maximize the binding affinity, known as the binding constant. This reduces the ability of the drug to dissociate from the binding site on the occurrence of various noncovalent interactions between amino acids at the binding sites with the hydrogen bonding (HB), electrostatic interactions, pi stacking, van der Waals interactions, and dipole–dipole interactions (Fig. 30.1).

Survismeter: Fundamentals, Concepts, and Applications
Man Singh and Sunita Singh

ISBN 978-981-4774-70-3 (Hardcover), 978-0-429-02761-1 (eBook)
www.panstanford.com

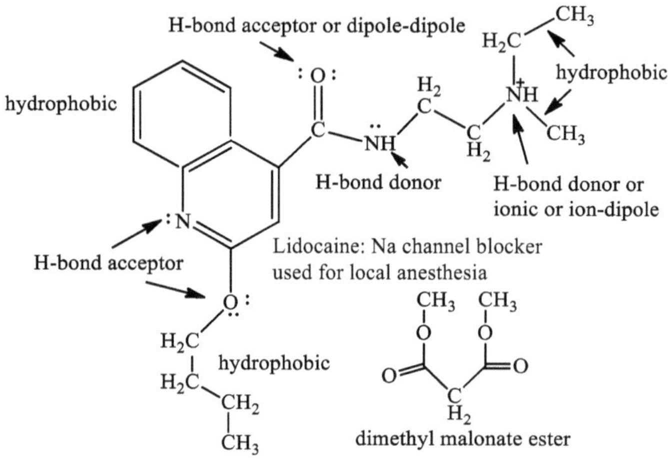

Figure 30.1 Hydrophilic and hydrophobic potential of lidocaine with adequate numbers of functional groups moderately influence shear stress and strain.

30.1 Entropy Production in Forming Micelles

The surfactant forms micelles when its concentration is greater than the critical micelle concentration (CMC) and the temperature is greater than the critical micelle temperature, or Krafft temperature. Friccohesity explains micelle formation by mutually balancing entropy and enthalpy. In water, hydrophobicity is a driving force for micelle formation despite unfavorable assembling of surfactant molecules in terms of both enthalpy and entropy. At a very low concentration of the surfactant, monomers are present in the solution. On increasing the surfactant concentration, a point comes at which unfavorable entropy production occurs out of clustering hydrophobicity tail. Such entropy production is overcome by a gain in entropy due to a release of solvation shells around the surfactant tail. So at this point, the hydrophobic tail gets segregated from water and starts forming micelles. Above CMC, the loss of entropy due to assembly of the surfactant molecules is less than the gain in entropy by setting free water molecules that were trapped in the solvation shell of surfactant monomers. Enthalpy works due to electrostatic interactions that occur between the charged parts of surfactants.

30.2 Micelle Packing Parameter

The following equation predicts the micelle or surfactant packing parameter (N_s) on self-assembly.

$$N_s = \frac{V_0}{A_{min} l_c}$$

Here, V_0 is the volume of the hydrophobic chain, l_c is the chain length, and A is the equilibrium area per molecule at the aggregate surface. N_s explains the shape of the surfactant. N_s depends on V_c. The cross-sectional area of the hydrophilic core of the aggregate is per molecule in the aggregate A and L_c.

The surface area (A) of the micelle could be determined from the surface excess concertation (τ, mol/cm^2) on conducting experiments for surface tension determination.

$$\tau = \frac{n_{total} - n^\alpha - n^\beta}{A}$$

α and β depict the phases, n^α and n^β depict the number moles in these phases, n_{total} depicts the total number of moles, and A depicts the surface area. For example, the prescribed moles of iodine are dissolved in the CCl_4 phase, which is brought in contact of the water phase. Partially, the iodine molecules' partition in the water phase is determined by titration or the interfacial tension (IFT) noted as under:

$$\gamma_{IFT} = P \quad \text{or} \quad \gamma_{IFT} = P = \frac{n_{I_2} \text{ in aqueous phase}}{n_{I_2} \text{ in } CCl_4 \text{ phase}}$$

Chapter 31

Block Copolymer Micelles

Micelles explain core-corona aggregates of small surfactant molecules and aggregates of amphiphilic block copolymers in selective solvents. A difference between aggregates of the shorter surfactant and the copolymer is in size. The building blocks of the copolymer are larger. The molecular weight of the surfactant is in the shorter range as compared to the weight of a block copolymer. Copolymers having larger hydrophilic and hydrophobic parts effectively express an amphiphilic nature as compared to surfactants. The building blocks of copolymer micelles behave like surfactants. The copolymer forms dynamic micelles, while the surfactants form kinetically frozen micelles. Such distinction could be measured with friccohesity. Most interestingly the trend of micelle formation of copolymers, surfactants, and dendrimers could be aligned with graphene, along with its functionalized products. In a nutshell such frontiers of graphene research could be focused as under:

31.1 Graphene Research

- New research interfaces in the areas of graphene functionalization with surfactants, salts, dendrimers, carbohydrates, and others

Survismeter: Fundamentals, Concepts, and Applications
Man Singh and Sunita Singh

ISBN 978-981-4774-70-3 (Hardcover), 978-0-429-02761-1 (eBook)
www.panstanford.com

- Research methodology for graphene functionalization in an ionic, oxidizing medium
- Functionalization and optimization mechanism of graphene with a functionalization agent
- Chemical as well as physical variables for structural stabilization
- Friccohesity of dipolar vectorization of functionalized graphene (FG)
- Supercapacitor ability of FG due to a lone pair of electrons on imidazole

The vectoral dipoles induce kinetic energy to monitor the molecular friccohesity and intramolecular multiple forces theory (IMMFT) of the molecules used in a specific quantity or fraction in liquids. Such vector and population both induce entropy and tentropy. So friccohesity with dipolar data could determine the density. Molecular dipoles act as an electric dipole with an inherent electric field, which is sharpened with a magnetic dipole and generates a magnetic field. Peter J. W. Debye was the first physical chemist to study molecular dipoles, and debye is the unit of dipole moment. A collection of N particles with charges q_i and position vectors $\mathbf{r}_i$ consists of electrons, with charge -e, and nuclei with charge eZ_i. Z_i is the atomic number of the ith nucleus. Dipole is an observable quantum mechanical dipole operator.

$$\mathbf{p} = \sum_{i=1}^{n} q_i \mathbf{r}_i$$

Here, **p** is valid for noncharged dipoles, with a total charge of zero. For a charged dipole the following equation is used:

$$\mathbf{p} = \sum_{i=1}^{n} q_i (r_i - r_c)$$

Here, r_c is the center of mass atoms of the molecule. For $+q$ and $-q$, two point charges, the electric dipole moment **p** is as under:

$$\mathbf{p} = q\mathbf{d}$$

The distance vector **d** points from the negative to positive charges, and the electric dipole moment vector **p** points from the negative to positive charges. Two-charge systems are electrical point dipoles consisting of two charges infinitesimally separated. A real inspiration is drawn from Robert Boyle's experiment of applying a partial pressure in a capillary to raise the liquid mixture in the capillary.

Chapter 32

Industrial Outlook of Cohesive Forces

Cohesive forces (CFs) prevent the solubilization of molecules in specific solvents and so facilitate phase extraction and separation, whereas friccohesity deals with the science of solvent-solvent interactions, solute-solvent interactions, etc. Cohesion or cohesive attraction or force is the action of mutual attractive forces and is an intrinsic property caused by the shape and structure of molecules. The CF facilitates the distribution of the electrons filled in an orbital. The distribution becomes irregular when the molecules come close to each other. This distance creates an electrical attraction that maintains a microscopic structure like a waterdrop. So cohesion allows for surface tension, creating a solid-like state where low-density materials could be placed. The higher the CF, the lower is the friccohesity; and the lower the viscosity, the higher is the conductance and the lower is the osmotic pressure.

Integrated physicochemical characterization becomes essential, and the device depicted in Fig. 32.1, noted as an oscosurvismeter, becomes helpful.

In such cases micelles are used to distribute the drug to all needed corners of the cells so the drug may enter the core of the micelles or the metal particle could enter the core. With time, the drug may drift out of the micelle core or the particle could come

Survismeter: Fundamentals, Concepts, and Applications
Man Singh and Sunita Singh

ISBN 978-981-4774-70-3 (Hardcover), 978-0-429-02761-1 (eBook)
www.panstanford.com

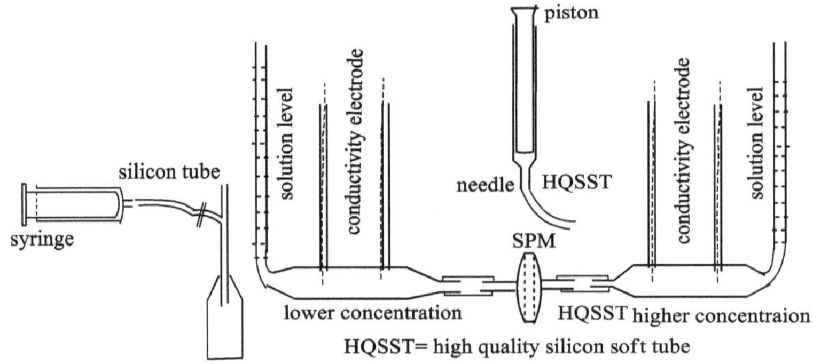

Figure 32.1 The structure and functional model of an oscosurvismeter.

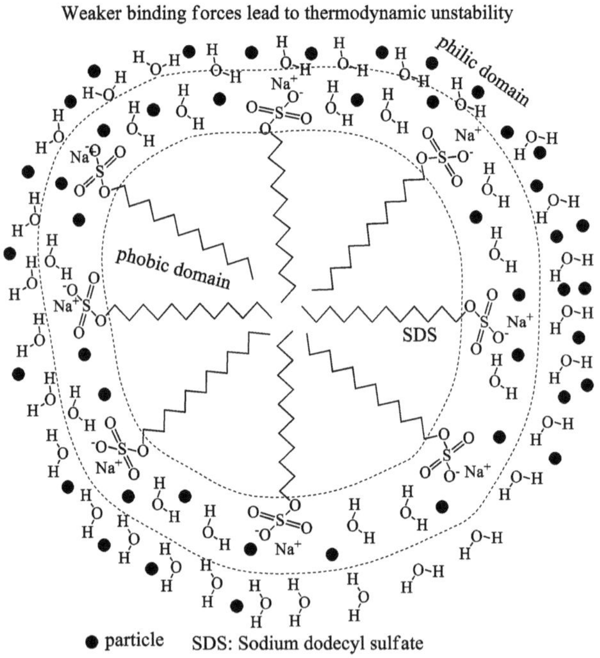

Figure 32.2 Constitutional frameworks of normal micelles with thin film of dipolar water at outer periphery.

Figure 32.3 Survismeter unit of oscosurvismeter to be used to measure surface tension and viscosity directly on dipping it in sample.

out from the philic core. This tendency is known as thermodynamic and kinetic unitability (Fig. 32.2). The particles that were tracked by the micelles and bound to philic layer of micelles through chemical forces may weaken with time due to Brownian motion caused by hydrophobic, longer alkyl chain of micelles. On weakening of the binding forces particle instability develops.

Such chemical processes could directly be measured without disturbing the micelles. For such measurements the UGPG Survismeter (UPS) model (Fig. 32.3) is used for surface tension and viscosity data.

The UPS model directly measures surface tension and viscosity. The UPS model does not contain the reservoir bulb (RB), a major limb. So it is directly dipped in the sample, which is filled in a beaker/conical flask. The remaining parts remain as such. It is most economical and appropriate for undergraduate and postgraduate students. In general the student fills the liquid sample in the RB of the survismeter but in the case of the UPS, it is directly dipped in the sample. So the use of the UPS eradicates several steps like washing and storage.

Steps for Measurement

(1) Clean the UPS and dry it.
(2) Mount the UPS on a stand vertically.
(3) Dip the lower tip of the UPS in the solution and lift the sample.
(4) Fix a reversible stopper in the pressure limb.

Viscosity Measurement

(1) Fix a stopper in the surface tension unit.
(2) Attach one silicon tube at one end of the L-shaped glass cone.
(3) Attach the other end of the silicon tube with a needle-fitted syringe.
(4) Fix the cone inside the socket of the UPS limb and ensure its airtight position.
(5) Reduce the air pressure inside the bulb of the viscosity unit to lift the liquid sample. The liquid sample is lifted to two-thirds of the buffer bulb (BB).

The UPS model is also most suitable for a study of drug release and binding study. For the study of drug release, the lower tip is attached with the syringe filter.

Chapter 33

Friccohesity of Walking Molecular Liquids and Nonwalking Molecular Liquids

Surfactants, drug molecules, and macromolecules with multi-interacting sites when dispersed in a solvent like water "walk" from one point to another. The surfactants initially are dispersed, and when their amount is increased, they start aggregating and finally form micelles. Monomer surfactant molecules keep moving from bulk phase to micelle and vice versa. Such molecules have alkyl chains of variable sizes, which act as legs or riders to move ahead in 1D, 2D, and 3D as per their kinetic energy and entanglement with the medium and cosurfactants. Such science becomes most fruitful for phase extractions, dispersion, thin film formation, dyeing, coating, heat dispersing, nanoparticle (NP) dispersion, and so on. Friccohesity is measured quantitatively using the survismeter.

The science and fundamentals of survismeter are of interdisciplinary nature and their understanding through questions and answers widen applicability of survismeter sciences. Thus few innovative and creative queries are presented along with their possible answers.

Survismeter: Fundamentals, Concepts, and Applications
Man Singh and Sunita Singh

ISBN 978-981-4774-70-3 (Hardcover), 978-0-429-02761-1 (eBook)
www.panstanford.com

(1) What is the survismeter, and what is its nomenclature?
Answer: It is laboratory equipment to measure various physicochemical properties. "Sur" stands for "surface"-related properties of molecules in liquid samples, and "vis" stands for "viscousness" or shear stress–related properties.

(2) How does the survismeter lead to the establishment of a new state of art in solution and nanoemulsion chemistry or sciences?
Answer: It tracks several physicochemical properties and the state of liquid mixtures.

(3) What is the basic invention of the survismeter?
Answer: The survismeter is based on Robert Boyle's partial pressure experiment applied in capillary which climbed up in the capillary itself.

(4) What is the reservoir bulb (RB) of the survismeter?
Answer: The RB of survismeter holds the sample safely and this sample is lifted to its measuring probes sententially for measurements.

(5) How does the survismeter determine its impact in solution chemistry and products formulation sciences?

(6) How does the survismeter determine hydration and micellization, which have the least entropy?

(7) How does the survismeter determine thermodynamic and kinetic stability of nanoemulsions and colloids?

(8) How does the survismeter determine coagulation and anticoagulating activities of liquid mixtures?

(9) What are working principles and theory of the survismeter?

(10) How does the survismeter supersede the existing devices in capillary sciences?

(11) How does Robert Boyle's low partial pressure work with the survismeter?

(12) How are the surface tension and viscosity interfaced?

(13) How does the survismeter distinguish a solution from a nanoemulsion?

(14) How does the viscosity increase and the surface tension decrease on decreasing the particle size?

(15) For a solution of smaller-sized particles, why is the viscosity higher and the surface tension and conductance lower?

(16) How does the survismeter determine aggregation, coagulation, and coalescence as dynamic processes?

(17) How does the survismeter determine solvation and emulsion formations, and what could be their friccohesity?

(18) In general, electrolytic salt solutions produce the highest surface tension and viscosity, contrary to nanoemulsions, which produce lower surface tension and higher viscosity and friccohesity both. Why do these systems behave in this manner?

(19) How does the survismeter determine molecular weights, and how does the survismeter determine the size of the ionic field in the case of aqueous salt solutions?

(20) How does the survismeter determine the degree of hydration of Li, Na, K, Rb, and Cs cations, with a rational approach?

(21) How does the survismeter measure van der Waals forces and London dispersive forces (LDFs)?

(22) How does the survismeter determine the Lennard-Jones potential (LJP) using salt hydrations?

Answer: For a NP capping study using phase hydration AuNPs are taken in the water phase and polyvinylpyrrolidone (PVP) is taken in the isopropyl alcohol (IPA) phase. Both phases are kept together to develop the liquid–liquid interface (LLI). The study could be repeated with alkanethiol or cysteine by replacing the IPA with them separately. The capping agent is taken in one liquid phase and the AuNPs in another. Both are kept together. Either the NPs or the capping agent moves toward capping so the kinetics of capping could be studied. Surfactants could be used for facilitating chemical species to move from one to another phase as per affinity.

(23) How does the survismeter accurately determine the phase extraction of a pollutant from one phase to another phase?

(24) How is iodine phase extraction from the CCl_4 phase to the water phase experiment conducted with the survismeter?

Answer: The phase extraction experiments are conducted by taking two sets of solute-solvent systems made in two immiscible solvents. For example, iodine is dissolved in CCl_4 and the I_2-CCl_4 solution is brought in contact with the water phase. The iodine starts moving from the CCl_4 phase to the water phase. In this process the interfacial tension (IFT) of the water–CCl_4 interfaces changes. Thus the IFT is accurate data that decides the phase extraction of iodine from CCl_4 to the water phase. Also petroleum ether and water could be used to perform similar experiments with drugs or dendrimers. Experiment of polymer swelling at the LLI of water and benzene, water and toluene, water and cotton seed oil. The PVP polyacrylamide could be used. These experiments could be performed in a systematic approach.

Ink Removing or Paint Removing Kinetics

This type of experimental science could lead to the development of ink removing or paint removing kinetics with suitable solvents. The ink phase or the paint phase could be kept in a suitable solvent with a suitable agent, and the friccohesity could be measured with time.

(25) Calculate Mansingh constant $M_c \cdot M_c = \sigma_0/(t_0 \cdot n_0)$. Here σ_0, t_0, and n_0 depict the friccohesity, viscous flow time, and pdn of the solvent, respectively.

(26) What are walking molecules, and how are they recognized by the survismeter?

Answer: Walking molecules are the surfactants, and they reduce the surface tension of the solvents or the liquids when these are measured using the survismeter.

(27) How is the lanthanide-loaded TTDMM (trimesoyl 1, 3, 5 tridimethyl malonate) ester dendrimer determined using the survismeter?

Answer: Since the TTDMM contains three similar units with delocalized electrons having a negative electron cloud centered at $-CH_2-(C{=}O)_2C$ of the malonate head group, the cation or the lanthanide cations get accommodated or conjugate in the void spaces of the TTDMM. This mechanism allows a substantial change in the friccohesity data as compared to the friccohesity of the unloaded lanthanide systems.

(28) How does the survismeter emerge to establish a new state of art in solutions and nanoemulsions?

Answer: The survismeter facilitates measurements of many friccohesic or physicochemical parameters together, saving chemicals, human efforts, and infrastructure in experimental sciences in areas of solution chemistry, nanoemulsions, biochemical, and biophysics. It tracks friccohesive engineering of binding and distribution activities of the molecules, which seem authentic data to study the behavior of drugs, catalysts, adsorbates, and antisolvation mechanism.

(29) What is the basic invention of the survismeter?

Answer: It develops an interlocutor mechanism to bridge cohesive forces (CFs) and frictional forces (FFs) of the solutes vis-à-vis solvents at the prescribed experimental conditions. It creates new interfaces of adhesive and anti-adhesive forces with respect to the CFs and FFs, respectively. The main breakthrough of the survismeter invention is the development of the pressure manipulating controlled pressure unit (CPU), which facilitates the sample to be bifurcated to other sample probes.

(30) What are the working principles and theory of the survismeter?

Answer: The working theory of the survismeter is pressure regulation and pressure reduction in the sample probes and the CPU. These in combination facilitate the many parameters with a single unit.

(31) How does the survismeter supersede the existing devices in what Leonardo da Vinci referred to as capillary sciences?

(32) How does Robert Boyle's law for partial pressure work with the survismeter?

(33) How are the surface tension and viscosity interfaced with the survismeter?

(34) Why is the survismeter a boon for nanosciences and technology?

Answer: The nanosciences directly deal with size-based physicochemical properties, and the survismeter working theory is measurement of materials in suitable solvents that develop bonding activities that affect the physicochemical properties. These are measured using the survismeter, and the variations therein could be interpreted to identify the science responsible for binding or dispersion activities of NPs in the selected medium at prescribed experimental conditions.

(35) Exp. no. 1: With the econoburette, a semimicro device, determine the strength of a given HCl sample with a standard sample of N/30 NaOH solution using phenolphthalein as the indicator.

(36) Exp. no. 2: With the econoburette, a semimicro device, standardize the given sample of NaOH with a standard N/30 oxalic acid solution as the primary standard using phenolphthalein as an indicator.

(37) Exp. no. 3: Determine the strength of a given $KMnO_4$ sample with a standard sample of oxalic acid.

(38) Exp. no. 4: With the econoburette, a semimicro device, titrate ferrous ammonium sulfate $[(FeSO_4)NH_4SO_4]$ with standard $K_2Cr_2O_7$ sample. The NaOH solution is standardized with a standard N/30 oxalic acid solution as the primary standard using phenolphthalein as the indicator.

(39) Exp. no. 1: Determine the surface tension and viscosity of aqueous 0.2, 0.6, 1.0, and 1.4 M NaCl electrolyte solutions together with the survismeter.

(40) Exp. no. 2: Determine the surface tension and viscosity of aqueous 0.2, 0.6, 1.0, and 1.4 M KCl electrolyte solutions together with the survismeter and compare the results with similar solutions of NaCl.

Note: Such differences determine the effects of Na^+ ($3s^1$) and K^+ ($4s^1$) cationic sizes and the strength of their ionic fields.

(41) Exp. no. 3: Determine the surface tension and viscosity of dimethyl sulfoxide (DMSO), a nonaqueous solvent, together with the survismeter and compare their data with those of water in similar experimental conditions.

(42) Exp. no. 4: Determine the surface tension and viscosity of aqueous 0.2%, 0.6%, 1.0%, and 1.4% soap solutions together with the survismeter.

(43) Exp. no. 5: Determine the surface tension and viscosity of aqueous 0.2%, 0.6%, 1.0%, and 1.4% detergent solutions together with the survismeter.

(44) Exp. no. 6: Determine the surface tension and viscosity of a petrol sample together with the survismeter and comment about the quality of the petrol sample.

(45) Exp. no. 7: Determine the surface tension and viscosity of diesel samples together with the survismeter and compare the data with those of the petrol sample.
(46) Exp. no. 8: Determine the surface tension and viscosity of ink samples together with the survismeter and comment about their quality.
(47) Exp. no. 9: Determine the surface tension and viscosity of aqueous 0.2, 0.6, 1.0, and 1.4 M glucose nonelectrolyte solutions together with the survismeter and compare with similar solutions of NaCl.
(48) Exp. no. 10: Determine the surface tension and viscosity of aqueous 0.02%, 0.06%, 0.10%, and 0.14% cetyl-trimethyl-ammonium bromide (CTAB), cationic surfactant solutions, together with the survismeter and compare with similar solutions of NaCl.
(49) Exp. no. 11: Determine the IFT of benzene and water immiscible phases using the survismeter.
(50) Exp. no. 12: Determine the impact of 0.02% CTAB on the IFT of benzene and water immiscible phases using the survismeter.
(51) Exp. no. 13: Determine the impact of 0.02% sodium dodecyl sulfate (SDS) on the IFT of benzene and water immiscible phases using the survismeter.
(52) Exp. no. 14: Determine the surface tension and viscosity of aqueous 0.002%, 0.006%, 0.010%, and 0.014% bovine serum albumin (BSA, a globular protein) solutions together with the survismeter.

Suggested Experiments for M.Sc. Practical Classes for Determining Physicochemical Properties of Substances using Survismeter, a Green and Clean Device

(53) Exp. no. 15: Determine the particle size of 0.002%, 0.006%, 0.010%, and 0.014% BSA solutions with the survismeter.
(54) Exp. no. 16: Determine the molecular weight of macro-molecules such as BSA using their 0.002%, 0.006%, 0.010%, and 0.014% solutions with the survismeter.
(55) Exp. no. 17: Determine the activation energy of a 0.002% aqueous BSA solution through viscosity measurements with the survismeter at 25°C, 30°C, 35°C, and 40°C temperatures.

(56) Exp. no. 18: Determine the critical micelle concentration (CMC) of CTAB and also determine the viscosity of the solution forming the CMC.
(57) Exp. no. 19: Determine the surface tension and viscosity of a 0.05% aqueous baby soap solution to be used as indicators for the safety of the tender tissues and skin of a baby.
(58) Exp. no. 20: Determine the surface tension and viscosity measurements of aqueous Tweens 80 and cetylpyridinium chloride (CPC) together with the survismeter for the estimation of emulsifying and surfactant agents, respectively.
(59) Exp. no. 21: Study the phase extraction of I_2 (iodine) from CCl_4 to water using the survismeter for friccohesity measurements.
(60) Exp. no. 22: Study the influence of Tweens 20 and 40 on the mutual wetting of hexane + water immiscible phases through IFT measurements with the survismeter.
(61) Exp. no. 23: Determine the hydrodynamic volume of macromolecules such as proteins like BSA, lysozyme, and egg albumin in aqueous solutions using the survismeter.
(62) Exp. no. 24: Determine the surface excess concentration (SEC) (τ, mol cm^{-2}) of cationic (CTAB), anionic (SDS), and nonionic (Tweens 20) surfactants using the survismeter.
(63) Exp. no. 25: Determine the contact angle theta (θ) with the survismeter.
(64) Exp. no. 26: Determine the column rise for the estimation of the wetting action of cationic, anionic, and nonionic surfactants.
(65) Exp. no. 28: Determine the density of a given sample with the column rise height h in the IFT capillary of the survismeter.
(66) Exp. No. 29: Determine the water-binding capacity with the size of cations of A1 group alkali metal salts HCl, LiCl, NaCl, KCl, RbCl, and CsCl series.
(67) Exp. No. 30: Determine the water-binding capacity with the size of halide anions of A1 group alkali metal salts HF, HCl, HBr, HI, LiF, LiCl, LiBr, and LiI and NaX, KX, RbX, and CsX series.

Chapter 34

Special Experiments with the Survismeter

Determine the impacts of trimesoyl 1, 3, 5 tridimethyl malonate (TTDMM), trimesoyl 1, 3, 5-tridiethyl malonate (TTDEM), and trimesoyl 1, 3, 5 tridipropyl malonate (TTDPM) dendrimers on the thermodynamics and kinetic stability of an oil (saturated)-water nanoemulsion obtained with a series of cationic surfactants.

34.1 Basic Concept of Micelles

Surfactants on dissolving pure water at a certain concertation develop stable micelles. Micelles occur when the cohesive forces (CFs) among their hydrophobic alkyl chain on developing hydrophobic-hydrophobic interactions and also the hydrophilic head on developing cohesion through water molecules attain critical stoichiometric ratios. Surfactant molecules when dispersed in less numbers are not in a position to develop micelles; this infers that adequate hydrophobic cohesion is missing at lower concentrations. But in adequate numbers they develop adequate cohesion against the friction forces (FFs) of the medium. Thus in micelles CFs as well as

Survismeter: Fundamentals, Concepts, and Applications
Man Singh and Sunita Singh

ISBN 978-981-4774-70-3 (Hardcover), 978-0-429-02761-1 (eBook)
www.panstanford.com

FFs participate, which generate a definite magnitude of friccohesity. The dendrimers due to their hydrophobicity induce certain changes, which are measured with friccohesity for adequate elaboration.

34.2 Theory for Nanoemulsions

Emulsions differ from solutions with respect to developing localized ionic forces with structure breaking and making chemical activities, contrary to the Brownian motions and higher friccohesity in the case of emulsions. Such distinctions become most authentic data for the study of nanoemulsions prepared with a series of cationic surfactants and monitored with a series of TTDMM, TTDEM, and TTDPM, the first-tier dendrimers. Fundamental studies to estimate their thermodynamic and kinetic stabilities could be made with their particles size, zeta potential, PDI, frequency factor of Arrhenius equation, quantitative intrinsic viscosity and surface tension, wettability, and friccohesity. A research methodology to find out stoichiometric ratios vis-à-vis physicochemical indicators could be evolved by conducting certain experiments with each dendrimer separately. The following systems are aimed for study.

- Oil-water-cationic surfactants
- Oil-water-cationic surfactants with variable cosurfactants
- Oil-water-cationic surfactant dendrimers
- Oil-water-cationic surfactant dendrimers with variable cosurfactants

Initially, Josiah Willard Gibbs (1839–1903) studied the surface stress as the amount of reversible work per unit area needed to elastically stretch a pre-existing surface. Instead of this definition, surface stress is defined as the amount of reversible work done per unit area needed to elastically stretch a pre-existing surface. A similar term is surface free energy, which represents the excess free energy per unit area needed to create a new surface. It is easily confused with surface stress. Although the surface stress and surface free energy of liquid–gas interfaces or liquid–liquid interfaces (LLIs) are the same, they are very different in the solid–gas or solid–solid interfaces, which would be discussed in detail later. Since both terms represent a force per unit length, they have been referred to as "surface tension," which contributes further to the confusion in the literature.

Chapter 35

Friccohesity of Dispersion or Flocculation of Colloidal Suspensions through the Surface Area of NPs' Zeta Potential

The friccohesity of dispersion or flocculation of colloidal suspensions through the surface area of nanoparticles' (NPs') zeta potential moderates the stability of the nanoemulsion. Adsorption of ions or surface-active molecules like dendrimers could promote charge stabilization via electrostatic repulsion through the Lennard-Jones concept. It could also be done by neutralizing the surface charge to destabilize dispersion. It depends on the cosurfactant that is adsorbed to the surface of curcumin (dispersed material). This science is centered on the cohesive forces (CFs) as well as frictional forces (FFs), or friccohesity. Curcumin is an antioxidant but is less soluble in aqueous solutions and hence its nanoemulsion is developed for better bioavailability. Friccohesity explains FFs and surface area for maintaining stability and instability. The mechanism for long-term stability or inducing rapid flocculation activities depend on molecular interaction engineering to widen the applications.

Many industrial processes like paper production, water treatment, and dewatering processes, require destabilization or

Survismeter: Fundamentals, Concepts, and Applications
Man Singh and Sunita Singh

ISBN 978-981-4774-70-3 (Hardcover), 978-0-429-02761-1 (eBook)
www.panstanford.com

flocculation of dispersed particulates. For foods, cosmetics, paints, and pharmaceuticals, the stabilization enhances the shelf life and consumer appeal. Many factors influence the stability in emulsions, dispersions, and suspensions. For example, adsorption of ions or surface-active molecules like dendrimers helps to promote charge stabilization through electrostatic repulsion under certain experimental conditions. But such arrangements could neutralize the surface charge and destabilize dispersions under different conditions. Also, depending on whether additives adsorb to the surface of the dispersed material or remain in the solution, they can promote stability or instability, respectively. Stability clarifies a solution to improve filterability or promote sedimentation or, alternatively, to prevent an emulsion from phase separation. Ionic dispersion for stability/instability is zeta potential. It controls the degree of electrostatic repulsion between neighboring particles. Friccohesity and zeta potential enhance stability and instability. They enhance stability or may induce rapid flocculation. Solubility and preferential solvation of flavonoids like naringenin occur in aqueous organic solvent mixtures. This needs rectification and justification of research work. The interacting abilities of chemical species in specific solvent systems resole solubility of flavonoids at prescribed experimental conditions. Preferential solvation is described as preferable composition of a solvent around a solute molecule in the solvation shell. For many pharmaceuticals it has not been studied so far. So explaining the mechanistic point of view of such phase incorporates vital interacting coordinates, such as friccohesity in the case of solute and solvent joint ventures.

So the equilibrium solubility of naringenin as an antioxidant agent could be determined in three pure solvents, water, acetonitrile, and 1, 4-dioxane, and in aqueous cosolvent systems, like water/ACN and water/DOX mixtures, using the UV-Vis spectrophotometric method. What has been a direct implication of solubility and interacting abilities in light of above and what has been actual hydrodynamics or solvodynamics workable in capillary flow sustainability (Fig. 35.1) could be tracked using the survismeter.

The preferential solvation of naringenin in binary mixtures is analyzed. These studies are essential to achieve further insight into the thermodynamic behavior of flavonoid naringenin solution and

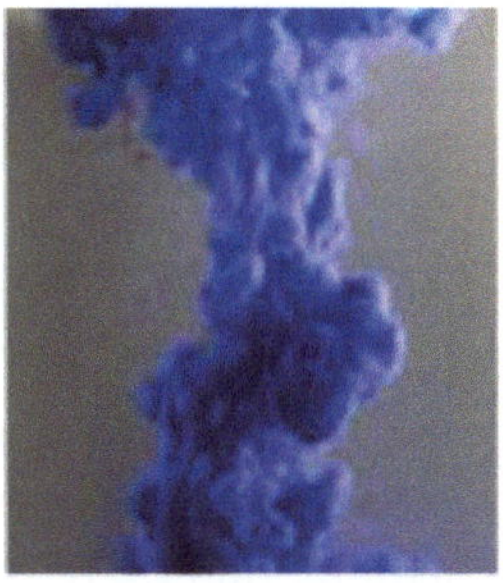

Figure 35.1 Colloidal state of liquid sample where the solute and solvent both have the moderate interactions that assists to hold the solvent around the surfaces of the solute molecule like flavonoids or curcumin noted as antioxidants through the van der Waals forces and orientational activities.

to understand the respective solute-solvent molecular interactions involved in the solubility and preferential solvation processes. So their structure making or breaking mechanism of said theme facilitate a binding-and-release dynamics similar to studies reported [6]. From linear fitting of data at the wavelength of maximum absorbance (ACN $\lambda_{max} = 285.2$ nm and DOX $\lambda_{max} = 284.4$ nm), the values determined for ε were 1.3850×104 and 1.3399×104 L $mol^{-1}cm^{-1}$ in cosolvents ACN and DOX, respectively. An order of solubility process is determined. The solvation free energy as interaction energies of solute with solvent is expressed as a summation of electrostatic interaction energy and van der Waals interaction between solute and solvent molecules. In polar water, the contribution of these interactions is very low for the dissolution process of less polar naringenin molecules. On adding ACN or DOX, the interactions among naringenin molecules are strengthened. The activities of molecular dimerization obtained through electrostatic structures dominate over solubilizing abilities of the resultant liquid mixture. The flavonoid could act as a Lewis base due to free electron pairs in oxygen atoms of –OH, >=O, and –O– groups to interact with hydrogen atoms in a protic solvent vis-à-vis role of pH as the pH influences the solubility dynamics. So a comprehensive and rationalized treatment to formulate a new model is needed. Medicines, drugs, ink, foods, clothes, and all living and nonliving objects are made up of the combinatorial degree of freedom out of Boltzmann energy distribution arrays of molecules and their interacting factors. Their entropies and free energy were studied by Rudolf Clausius and Helmholtz Gibbs, respectively. Their theories lead to the generation of stable geometrical structures from the electronic and atomic dimensional and dynamic molecular framework, which has attracted introspection regarding, elucidation of, and quantization of molecules. Thereby, fundamentals of synthesis and characterization of molecules correlate the FFs of solute that fractionate the CFs of medium. Had there been no interaction with the medium, there would have not been any homogenization, and distribution of CFs to be scattered with higher networking. For homogenization process, the viscosity increases and surface energy decreases with highest friccohesity.

Chapter 36

Critical Insight into Molecular Activities for Tentropy and IMMFT

Molecules are made up of atoms, and atoms have electrons, protons, neutrons, and other small particles. The combinations of atoms for making molecules are monitored by several theories, such as Born–Oppenheimer approximation (BOA), where the electrons and protons develop several motions to establish a stable chemical bond for a molecule. The proton gets settled at the center because it is slightly heavier in weight than electrons. The dynamic, kinetic, and mechanical interacting models $(-e^2/r)$ depend on the electrostatic energy and kinetic energy $(p^2/2\text{m})$. The molecular activities in specified liquid mixtures like food indigestion formulate properties of resultant materials. These activities of molecules generate their activities which are defined by intramolecular multiple forces theory (IMMFT). The intramolecular activities within a single molecules generate several domain that define entropic ensembles causing a tentropy. So a variety of molecules, like proteins, enzymes, hormones, carbohydrates, and salts, that ionize and several others remain active during working time and in the working state, developing the chemical potential to sustain life. A few molecules, such as proteins, supramolecules, and dendrimers, develop residual

Survismeter: Fundamentals, Concepts, and Applications
Man Singh and Sunita Singh

ISBN 978-981-4774-70-3 (Hardcover), 978-0-429-02761-1 (eBook)
www.panstanford.com

forces and flow with a certain unique behavior in our circulatory systems. For example, cholesterol, being a sluggish (comparatively saturated) molecule, gets deposited in our arteries and causes atherosclerosis. Thus, molecules such as phospholipids, oils, and fats cause frictional forces (FFs), cohesive forces (CFs), and adhesive forces with different levels of wetting and are not dissolved or do not even get wet out with the water. So wetting is essential to ensure healthy working of our bodies. In general, viscosity for FFs, surface tension for both adhesive forces and CFs, wetting coefficient, and interfacial tension (IFT) between lipid and water phases are being studied individually, but the survismeter determines all these parameters together. The contact of the fat or cholesterol is restricted with the aqueous phase. The survismeter also measures the contact angle with the following equation:

$$\frac{h\rho gr}{2\gamma} = \cos\theta$$

Here, θ is the contact angle, h is the height of column rise in the capillary, ρ is the density, g is the gravitational force, r is the inner diameter (id) of the capillary, and γ is the surface tension. Similarly, the wetting coefficient is determined by the following equation:

$$\frac{\eta h^2}{t} = w_c^0 + A\gamma$$

Here, η is the viscosity, γ is the surface tension, h is the column rise in time t, w_c^0 is the wetting coefficient, A is a constant, and γ is the surface tension. The contact angle theta and the wetting coefficient are inversely promotional to each other, listed as under:

$$\theta = \frac{k}{w_c^0}$$

Here, k is the proportionality constant and the other symbols represent the usual. If the liquid does not develop a meniscus then $\theta = 0$, where $\cos\theta = 1$. The surface tension is determined by the static method using the survismeter; r is constant. The inward pull (meniscus) is the surface tension that develops some angle theta. The molecules that cause much surface pull develop a higher surface tension because stronger forces work on the surface to pull down the surface molecules. When less pull is applied then weaker molecular forces work on the surface and the theta contact angle has a low value. The adhesive force being applied on the surface area of the glass part matters a lot where the contact angle is concerned.

The wetting activities initiate the quantized approach on integration of cohesion and spreading forces. A weakening of cohesion of similar molecules allow higher wetting. The quantum chemistry of nanoemulsions is studied by monodispersing particles by computing their energy. The wave function of an average-sized molecule is explained by BOA, named for Max Born and J. Robert Oppenheimer. For example, a benzene molecule contains 6 nuclei of C and 6 nuclei of H, with a total of 42 electrons. Therefore, a time-independent Schrödinger equation has determined the energy and wave function both via a partial differential eigenvalue equation considering the spatial coordinates of *electrons* and *nuclei*. The BOA enables computation of wave function in two less complicated consecutive steps, proposed in 1927 by Born and Oppenheimer. It is still indispensable in quantum chemistry. The wave function of a molecule on the basis of its electronic and nuclear (vibrational, rotational) components is proposed as under:

$$\psi_{\text{total}} = \psi_{\text{electronic}} \times \psi_{\text{nuclear}}$$

Initially, the BOA solves the electronic Schrödinger equation to resolve the wave function $\psi_{\text{electronic}}$ on the basis of electrons where the nuclei are fixed in equilibrium configuration. For the second step, $\psi_{\text{electronic}}$ serves as a potential in the Schrödinger equation—containing only the nuclei of benzene—an equation with 36 variables. The success of BOA is credited to a high ratio between nuclear and electronic masses. The BOA in quantum chemistry resolves solutions for H_2, which are the lightest molecules, and could be extended for larger molecules. The electronic energies generated from nuclear potential are kinetic energies, interelectronic repulsions, and electron–nucleus attractions. These activities lay a foundation of tentropy and IMMFT both. The nuclear potential averages the electron-nuclear attraction. In molecular spectroscopy the ratios of periods of electronic, vibrational, and rotational energies are related to each other on scales on the order of a thousand. The nuclear spin energy is very small and is omitted. BOA predicts energy components as under:

$$E_{\text{total}} = E_{\text{electronic}} + E_{\text{nuclear}} + E_{\text{rotational}} + E_{\text{nuclear}} + E_{\text{friccohesity}}$$

Chapter 37

Oil Oxygen Capturing Ability of Molecules Studied with the Survismeter

The American Oil Chemists' Society (AOCS) approves the oil stability index (OSI), which determines the relative resistance of fat and oil to oxidation. The oxidation of oil is inhibited if its active oxygen is replaced, which is noted as the active oxygen method (AOM). It is done in intensive and the required chlorinated solvents. Both fats and oils undergo oxidation, but the rapidity of the oxidation depends on the degree of unsaturation, antioxidants, and storage conditions. The rate of oxidation in OSI analysis remains slow until the resistance to oxidation is removed. This time slag is known as the oxidation induction period. After the induction period, the rate of oxidation increases dramatically. Unlike peroxide value the free fatty acid analysis for good or bad oil is at a particular time, and OSI analysis has a predictive value. The OSI analysis predicts stability of oils and also evaluates the effectiveness of the antioxidants. It also determines how much longer frying oil can be used before it goes bad. Soybean, palm, peanut, sunflower, corn, coconut, and canola plant oils are analyzed. The OSI time varies from less than 2 h to over 100 h. Friccohesity is the most authentic and green method to determine the stability of oils using the survismeter.

Survismeter: Fundamentals, Concepts, and Applications
Man Singh and Sunita Singh

ISBN 978-981-4774-70-3 (Hardcover), 978-0-429-02761-1 (eBook)
www.panstanford.com

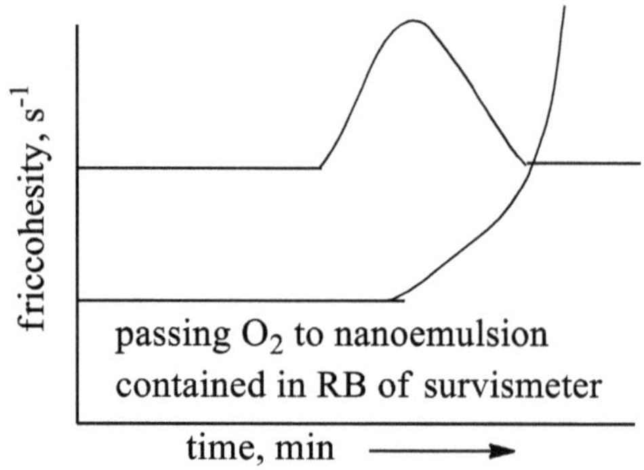

Figure 37.1 Reflects resultant state of nanoemulsion and its interacting activities with the O_2 gas molecules, passed in it. Friccohesity varies with the time on adsorbing the O_2 where an increasing friccohesity infers a substantial response of nanoemulsion with O_2 where it gets adsorbed and weaken the cohesive forces but an increase and decrease in friccohesity predicts an adsorption of O_2 which could lead to make the clusters with the stronger cohesive forces so the friccohesity decreases.

37.1 Experimentation for Oil Stability Index

Oil or melted fat (5.0 g) is weighed into a disposable glass test tube and is placed in a heating block at 110°C. Clean, dry air is bubbled through the sample, and the effluent stream of air is bubbled through a collection tube filled with ultrapure water. An electrode is placed in water to monitor the conductivity. As the oil oxidizes, volatile organic acids are given off—getting trapped in a collection tube—and increase the conductivity or decrease the friccohesity of water. The instrument generates a plot of conductivity versus time and determines an inflection point in the conductivity curve. This inflection point is defined as OSI time, and a conversion is used to convert the OSI time into a corresponding active oxygen method time. Friccohesity does not need such simulations and directly reflects the stability of the oil samples (Fig. 37.1).

Compounds are formed, and there is a free flow of electrons. These electrons come in contact and exchange needful energy. This activity is noted as activation energy. For higher yields new routes are needed. New, encouraging environments and openings are to be explored for new knowledge generation. If expert peers are together then their expert expression opens and ignites new hopes that create new establishments. So we have to promote creative ideas, focusing on the following:

- Creative research culture inspiring new and novel knowledge creation
- Involvement of universities, as centres of excellence, to generate ideas for new state of theb arts by adopting out-of-the-box thinking because innovative activities establish logic and balance for conducting path-breaking experiments
- Philosophy leading to constructive interference
- Raman stokes and antistokes
- Leonardo da Vinci and 16th century renaissance
- Robert Boyle's experiment
- Temperature disrupting hydrogen bonding (HB), studied using the survismeter

37.2 Friccohesity and Structural Application

The idea for developing new structural frameworks using platinum (+4) with a series of benzylamine derivatives has been formulated, and platinum(IV) complexes $[PtCl_4(BAD)_2]$ have been synthesized. These molecules have been used to study intercalation of DNA, and some of them, especially bis((4-chlorophenyl) methanamine) tetrachloroplatinum (M4CBA), have been found most effective. The synthesis of bis(phenylmethanamine) tetrachloroplates and others was done using a general procedure and characterized by nuclear magnetic resonance (NMR), Fourier-transform infrared (FTIR) spectroscopy, thermogravimetric analysis (TGA), differential scanning calorimetry (DSC), liquid chromatography–mass spectroscopy (LC-MS), and others.

On the basis of the results for DNA intercalation study, the drug-friccohesity-interaction (DFI) model is proposed. Their studies could be conducted with several other molecules, like dendrimers, proteins, surfactants, complexes, metallic salts, organic salts, and natural products, for finding their applications through their

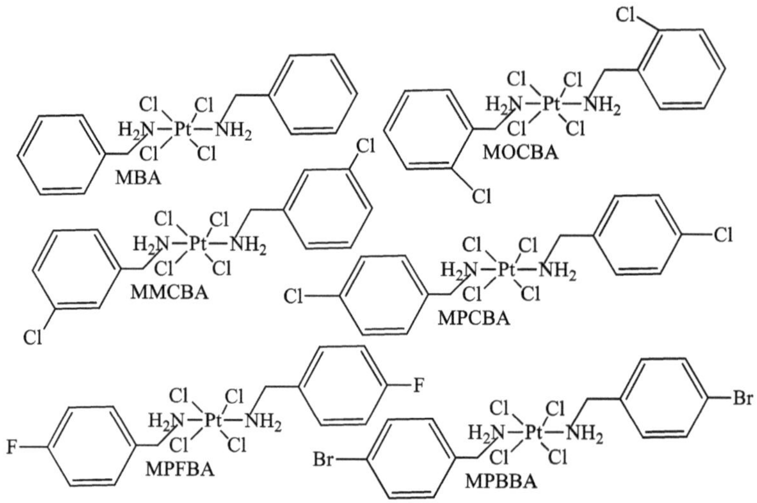

Figure 37.2 Series of MBA [bis(phenylmethanamine) tetrachloroplatinum], MOCBA [(bis-orthochlorophenyl methanamine) tetrachloroplatinum], MMCBA [bis(meta-chlrorومethanamine) tetrachloroplatinum], MPCBA [bis(para-chlroromethanamine) tetrachloroplatinum, MPFBA [bis(para-fluoro-methanamine) tetrachloroplatinum], MPBBA [bis(para-bromo-methanamine) tetrachloroplatinum]. The ortho (O), meta (M), and para (P) depicts positions of halide atoms so MOCBA, MMCBA, MPCBA, MPFBA, and MPBBA also noted as M2CBA, M3CBA, M4CBA, M4FBA, and M4BBA, respectively.

physicochemical properties (Fig. 37.2). A few experiments could be designed, as under:

Physicochemical dynamics of dendrimers with surfactants in aqueous medium lead to unique molecular interaction engineering that results in the formulation of valuable industrial liquid mixtures. Such liquid mixtures can act as similar ionic liquids with stronger solubilizing abilities.

Molecular signatures, translation, and expression are tracked with the survismeter.

Chapter 38

Monodisperse Nanomaterials for Drug Delivery Systems

Molecular expressions in liquid mixtures at variable pH, additives, and bis(phenylmethanamine) tetrachloroplates are tracked by friccohesity, and molecular expressions in solid state are tracked with Fourier-transform infrared (FTIR) spectroscopy.

The interaction activities of molecules as per their electronic configuration, functional groups, and basic structure become fundamental for formulating functional liquid mixtures of desired polarity and heat-holding ability. Therefore, various structural optimizations are done to choose different molecular structures noted as below:

- Study of solution and nanoemulsions with bis(phenylmethanamine) tetrachloroplates
- Electrostatic dynamics and Lennard-Jones potential (LJP) for the capping of nanoparticles (NPs)
- Monopolar and dipolar interactions of benzene salt or monopole-pi-interaction
- Van der Waals forces and London dispersive forces (LDFs) for particle distribution
- Superhydrophobe trimesoyl tridihexyl malonate ester (TTHM) for holding water for capturing ions to purify a water sample

Survismeter: Fundamentals, Concepts, and Applications
Man Singh and Sunita Singh

ISBN 978-981-4774-70-3 (Hardcover), 978-0-429-02761-1 (eBook)
www.panstanford.com

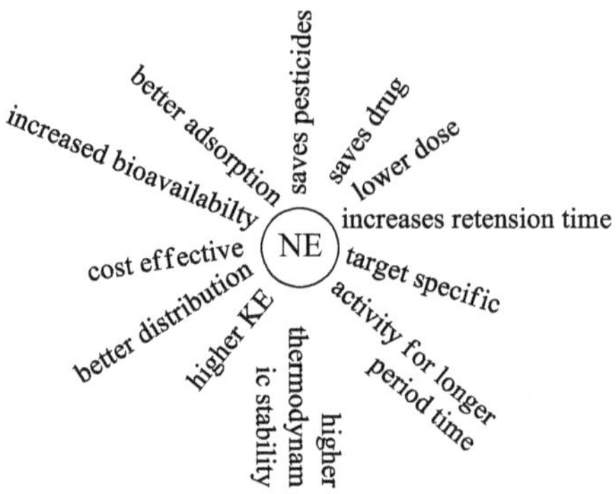

Figure 38.1 Wider application spectra of nanoemulsions with their specific potential.

The water hydrates in marshes that catch fires could be due to hydrocarbons like glucose, proteins, and biocatalysts. Such water-hydrating activities of hydrocarbons develop a unique interacting nature. The disease due to molecular disturbances is caused by the structural disturbances of proteins because the electronically charged molecular environment induces interaction where water, minerals, proteins, or hormones are engaged. Molecular combinations and their physical states develop valuable physicochemical processes, for example, ice, water, and vapor exist in different geometries, have different activities, and cause different impacts. So the whole body of the molecules undergoes reorientational changes through its electrostatic poles. However, nonmolecular materials do not speak much. Such systems may act as effective sensors by undergoing infinitesimal changes in their structures. The metal mercury as a liquid is more sensitive for temperature. Lord Kelvin, Torricelli, and others developed their temperature sensor and pressure by making the best use of such properties of chemical substances. Scientists are tracking molecular language, which has helped in developing various devices for measuring MP, BP, and FP. Each molecular substance has fixed intrinsic properties. Therefore, friccohesity as a sensor indicates the state of various molecules along with their responses toward selected solvents, buffers, or other media used for formulating the products.

38.1 Salts Have an Effect on Double Activities of Tweens

Tweens from 20 to 80 develop micellar solutions, but NaCl salts transform them into emulsions because the salts develop hydration spheres with water and in the process the tweens drift away slightly and coagulate, developing opaque solutions. The NaCl decreases the surface tension of the aqueous-tween solution whereas it increases the surface tension of the water where the tweens are absent. This proves that tweens with many -OH groups disrupt the hydration spheres developed by the ions of NaCl and the cohesive forces (CFs) that were developed due to cationic and anionic hydration spheres in the absence of tweens are disrupted now, when tweens are added with aqueous NaCl. NaCl decreases the surface tension of aqueous Tw60 by the 0.92 mN/m but the decrease is by 1.05 mN/m of aqueous Tw80, which produces a critical change in the resultant mixture. Tw80 remains a clear solution, but Tw60 becomes turbid due to coagulation with NaCl. It seems that Tw60 structurally affects the hydration spheres of Na^+ and of Cl^- due to the presence of many -OH groups but Tw80, with longer alkyl chains, does not have the same effect. The ST stands for surface tension in Fig. 38.2.

Tw60 (Fig. 38.3) moves toward phase separation and gives an opaque solution, with a larger gap between the water–surfactant interface on a nanometer scale.

Tw80 (Fig. 38.4) does not move toward phase separation with NaCl and gives a clear solution, with a smaller gap between the water–surfactant interface on a nanometer scale.

38.1.1 Surfactant and the Least Surface Tension

Alcohols produce a peculiar trend of surface tension values with aqueous glycerol mixtures. Their expressions are given in Figs. 38.3 and 38.4, and their data are plotted in Fig. 38.5. The ethanol acts as a surfactant because it has produced the least surface tension, but methanol, with the shortest alkyl chain, has produced a higher surface tension. The size compatibility of ethanol with glycerol could be the reason for such behavior, but propanol also behaves

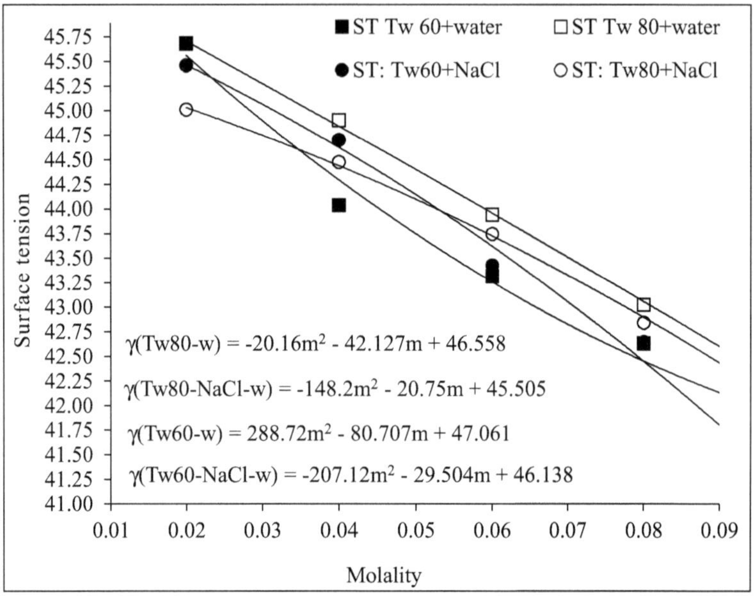

Figure 38.2 Surface tension of polysorbates in their aqueous solutions with their increasing compositions.

as a surfactant. Figure 38.5 is very significant as it depicts the determination of the hydrocarbon chain. It is seen that the medium is glycerol for methanol, ethanol, and propanol alcohols in similar experimental conditions but methanol produces a slightly higher surface tension. The overall trend is methanol > propanol > ethanol, which proves that the surface tension depicts the length of the alkyl chain. Methanol probably acts as a similar unit on forming hydrogen bonds with glycerol. However, ethanol, with one $-CH_2-$ (methylene) group, could strongly be disrupting the surface tension of glycerol. In this context propanol, with a longer alkyl chain, weakens the hydrogen bonding (HB) because of the induction effect and hence produces a slightly higher surface tension than ethanol. However, the study needs verification by taking alcohols with longer alkyl chains. The length of the alkyl chain affects hydrophobicity due to similar electron clouds. It seems that CH_3OH develops interaction with glycerol in a ratio of 1:3. Three molecules of methanol interact with one molecule of glycerol and induce a hydrophobic environment,

Figure 38.3 Tween 80 (Tw80) with a single double bond in it its alkyl chain develops nanoemulsions contrary to the Tw60 which without any double bond develops microemulsions. Tw80 and 60 distinguish the role of double bonds in dispersing the tweens in water where the Tw60 has higher hydrophobicity and develops stronger hydrophobic-hydrophobic interactions that induce cluster formations but the Tw80 due to a double bond develops stronger hydrophobic interactions.

Figure 38.4 The effect of ionic fields of the Na^+ and Cl^- ions which develop their ionic hydrations are seen to reorient the state of hydrophobic–hydrophobic and hydrophilic–hydrophilic interactions and domains. The NaCl salt has structurally affected the nanoemulsions of Tw 80 and microemulsion caused by Tw60 respectively due to engaging the water molecules in the cluster formation and ionic hydration formation that inhibit the water molecules to move the micelles formation.

which increases the influence of the surfactant due to the CFs (Fig. 38.6).

The mechanical, chemical, electronic, and oscillatory motions have the most substantial influence on the molecules being permitted in a prescribed friccohesic capillary. When the molecules are exposed to the electrical forces, magnetic forces, and laser then the molecules undergo translational, structural orientations, which

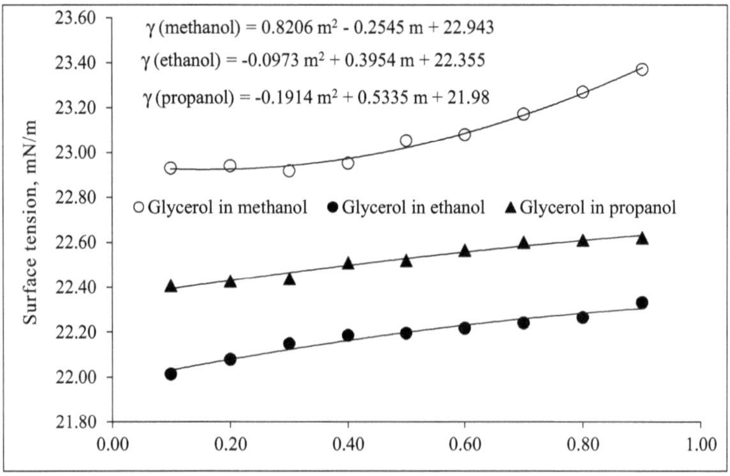

Figure 38.5 Surface tension of glycerol in series of alcohols distinguishes a contribution of length of alkyl chain of alcohols.

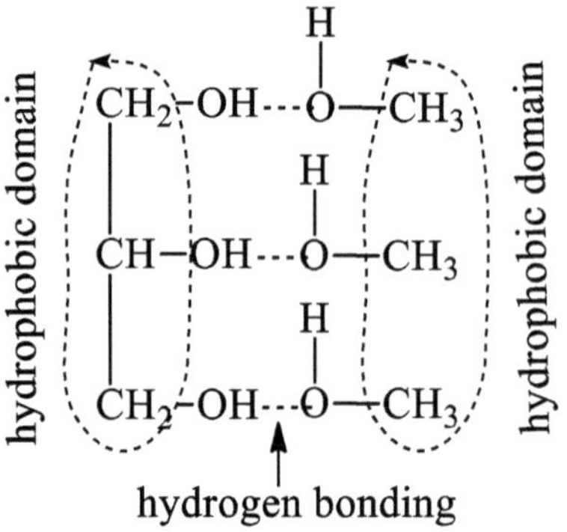

Figure 38.6 The probable mechanism of hydrogen bonding of glycerol molecules responsible for expressing exceptionally higher shear stress.

could become a substantial probe for understanding the interaction behavior. Figures 38.7(a,b) depict frequency and wavelength relations for developing a magnitude of energy.

The energy of electromagnetic radiation if subjected to the sample during flow in the friccohesic capillary could be calculated using Hook's law, as under:

$$\nu = \frac{1}{2\pi}\sqrt{\frac{k}{\mu}}$$

Here, ν is the stretching frequency, k is Hook's or spring or bond constant, and μ is reduced mass.

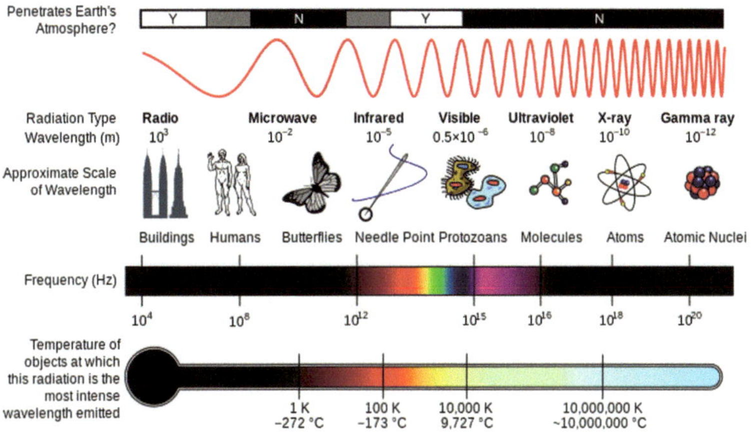

Figure 38.7 State of energy of light waves expressed as wavelength for inducing light and matter interactions causing isotropic and aniostropic (tentropic) activities within a single molecule.

38.2 Viscosity Enhancement

The viscosities of tetramethylammonium hydroxide (TMAH) (Fig. 38.8) in aqueous TMAH and in 0.5%, 1%, and aqueous 2% carboxymethylcellulose have been determined with the survismeter

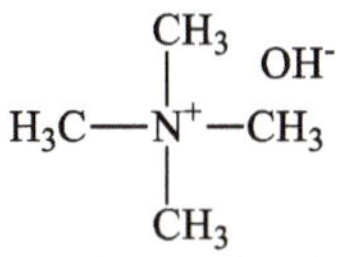

Tetramethylammonium hydroxide

Figure 38.8 Highly localized hydrophobicity with highly electron releasing activities toward quaternary nitrogen to be used as additive.

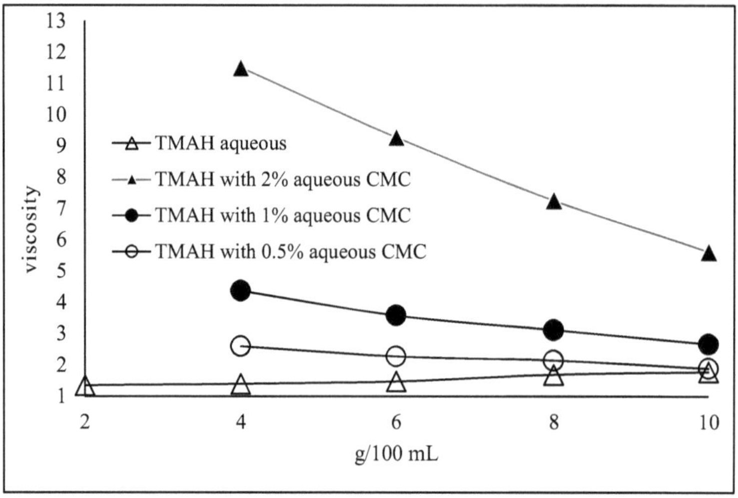

Figure 38.9 Viscosity of aqueous tetramethylammonium hydroxide with carboxymethylcellulose.

and are plotted in Fig. 38.9. The viscosities increase with carboxymethylcellulose percentage, where the 2% carboxymethylcellulose has produced nonlinear increments with TMAH g/mol whereas the 0.5% carboxymethylcellulose has increased viscosities almost linearly. The nonlinear increase depicts carboxymethylcellulose-carboxymethylcellulose over TMAH-carboxymethylcellulose interactions, whereas the linear increase indicates a dominance of TMAH-carboxymethylcellulose interactions.

Contrary to the viscosities of TMAH-carboxymethylcellulose systems (Fig. 38.9) the viscosities of melamine-formaldehyde-polyvinylpyrrolidone (MFP) polymer resin (Fig. 38.10) show exception trends of viscosities (Fig. 38.11), which proves that the fluid dynamics of the MFP resin is of multiple orders. At lower concentrations the values are higher, at 0.0047% MFP the value is the lowest, but 0.0047 onward the values increase. An increase depicts stronger interactions, but lower values indicate weaker water–MFP resin interactions [7].

Also the viscosities of poly-*N*-vinyl pyrrolidone oxime (PVPO) and poly-*N*-vinyl pyrrolidone oximo-L-valyl-siliconate (POVS) were studied (Fig. 38.12), where the viscosities increased with PVPO

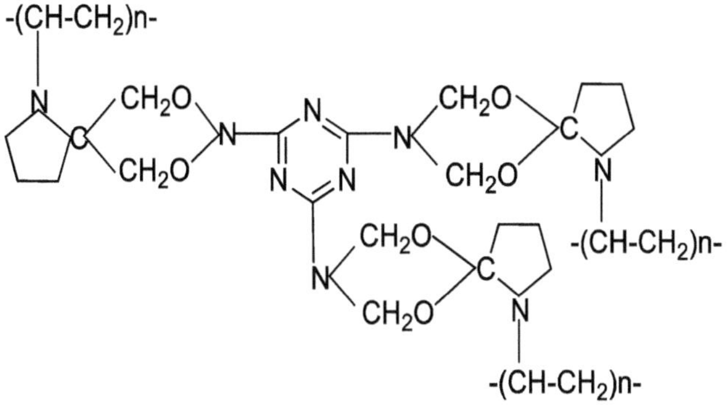

Figure 38.10 Melamine-formaldehyde-polyvinylpyrrolidone (MFP) Resin.

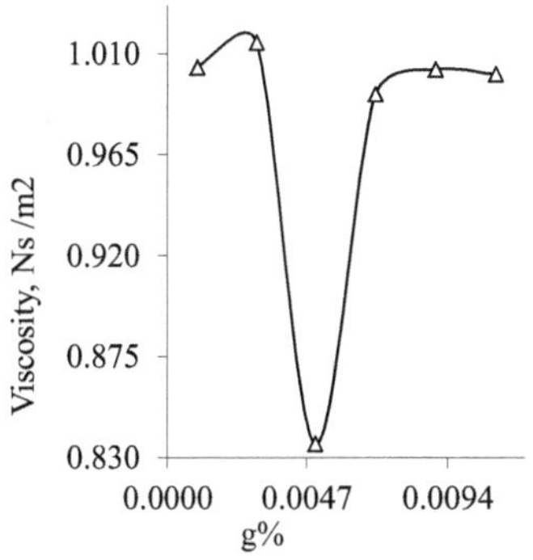

Figure 38.11 Unique shear stress for aqueous MFP resin solution.

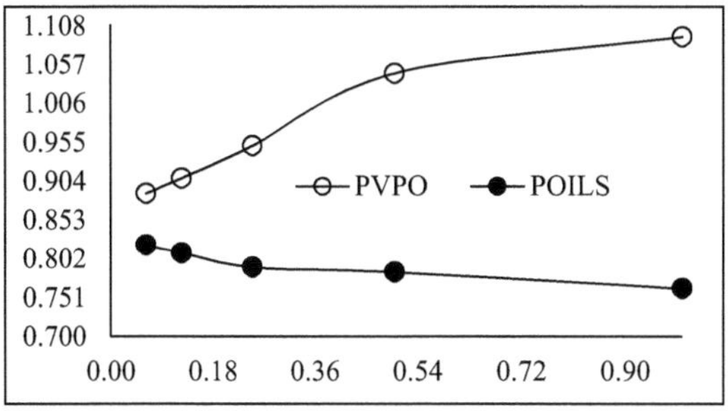

Figure 38.12 Viscosity values with compositions of PVPO and POILS.

compositions but decreased for poly-*N*-vinylpyrrolidone oximo isolucine-siliconate (POILS) or POVS. The oxime group of PVPO has elongated the water-engaging capacity but L-valyl siliconate has decreased the values, which may be due to a symmetric forward rolling out of the POILS molecules in water. So it proves that POVS developed weaker interactions with water.

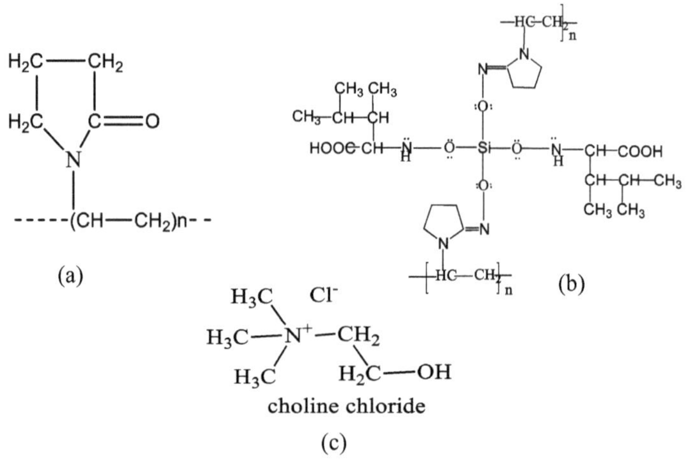

Figure 38.13 Structures of (a) PVPO, (b) PVOS, and (c) choline chloride.

The viscosities of MFP, PVOH, and PVOS with choline chloride could help develop a new understanding of their interactions with choline. Since a variety of molecular structures generate variable responses of the structure with the solvent used as the medium, such measurements could be useful for biologically active molecules.

Chapter 39

Radius of a POVS Macromolecule

The particle size of a poly-N-vinyl pyrrolidone oximo-L-valyl-siliconate (POVS) macromolecule is determined and is plotted in Fig. 39.1. Here the size is normalized when an adequate number of POVS macromolecules are stabilized.

$$\varphi = \frac{4}{3}\pi r^3 N_A c, \; N_A = 6.023 \times 10^{23}$$

A comparatively higher population of POVS becomes an almost straight line because the coalescence gets saturated and as per its molecules becomes stabilized. Especially the molecular interaction leads to the generation of an element of cohesive forces (CFs) with a definite friccohesity, and thus the density data along with fluid dynamics become significant. Sometimes, the molecules show variable behavior so several physicochemical properties (PCPs) are measured for better explanations of their interaction, which help in designing better applications. Therefore, for the study of such molecules, a new device is developed, known as the depsurvismeter, depicted in Fig. 39.2. The depsurvismeter basically measures the density, potential, surface tension, and viscosity of liquid samples together for a better and clear understanding of the dissolution or dispersion of molecules like drugs or the dispersion of dendrimers.

Survismeter: Fundamentals, Concepts, and Applications
Man Singh and Sunita Singh

ISBN 978-981-4774-70-3 (Hardcover), 978-0-429-02761-1 (eBook)
www.panstanford.com

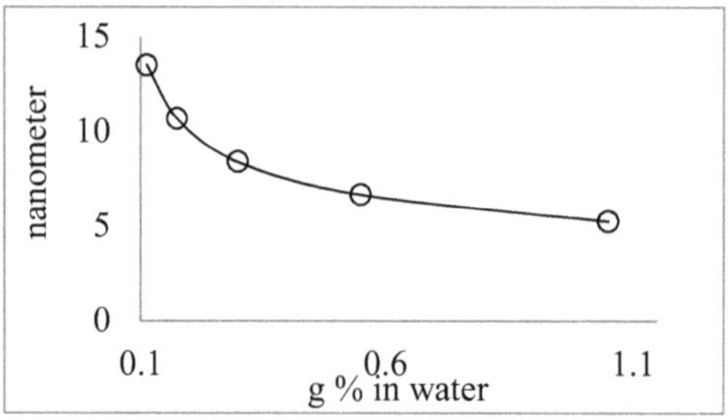

Figure 39.1 Particle size of molecules with a slightly more variation at lower concentration due to much solvent binding.

39.1 The Description of a Depsurvismeter

The depsurvismeter has a glass float (GF) inserted in the liquid sample filled in the reservoir bulb (RB) B1. S1, S2, S3, S4, and S5 are sockets of the respective units, B2 is the controlled pressure unit (CPU), B3 and B4 are bulbs for the surface tension unit, and B5 and B6 are bulbs for the viscosity unit. The labels m1 to m3 and m4 to m6 are marks on the viscosity and surface tension units, respectively. T1 and T2 are capillary tubes to support the electric leads, and the number 13 depicts the major limb (ML) opening in B1. E1 and E2 are silicon connectors to be attached with a piston pump. The potentiometer is attached through electric leads or the wires as a connector to pass the potential to the potentiometer. The platinum foil is used as an electrode, and the scales are shown in an expanded mode for a higher resolution to read the moment of the float. The depsurvismeter determines the density of nanoemulsions and other similar fluids.

39.2 Nanoemulsions and Allied Features

Nanoemulsions are one of the most efficient dispersed nanosystems of droplets within the nanometer size and are thermodynamically and kinetically stable transparent or translucent dispersions of oil and water. These are stabilized by the interfacial linkages established by the surfactants and cosurfactants of droplet size

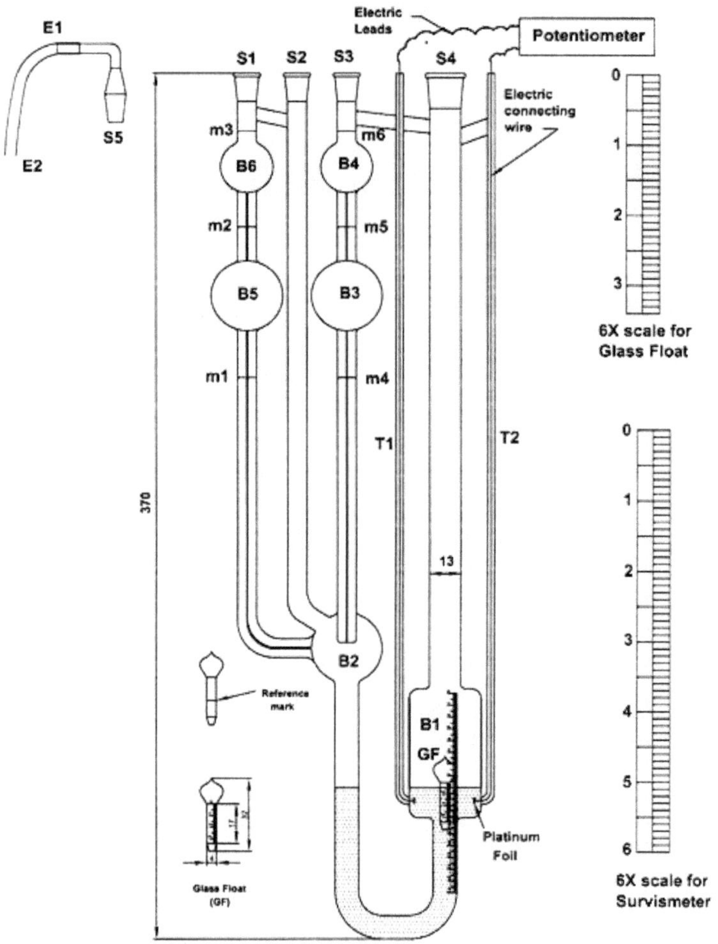

Figure 39.2 The diagram of a depsurvismeter holding magnetic float in bulb B1.

<100 nm. A nanoemulsion assists drug carriers in the delivery of therapeutic agents. Nanoemulsions have a fine core and accommodate drugs in the monodisperse mode with weak van der Waals forces. Nanoemulsions are transparent contrary to microemulsions (Fig. 39.3). Techniques and parameters like microfluidization, high-pressure homogenization, low-energy emulsification, and solvent

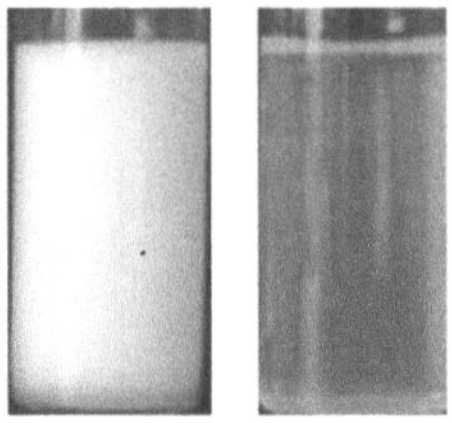

Figure 39.3 The physical appearances of microemulsions and nanoemulsions based on the sizes of the particles that reflect, diffract, and refract visible light, developing color.

evaporation are used for nanoemulsion characterization like droplet size analysis, viscosity determination, drug content, refractive index, pH, zeta potential, thermal stability, release, and in vitro skin permeation study. Three types of nanoemulsions are formed depending on the composition.

- Oil-in-water nanoemulsions, wherein oil droplets are dispersed in a continuous aqueous phase
- Water-in-oil nanoemulsions, wherein water droplets are dispersed in a continuous oil phase
- Bicontinuous nanoemulsions, wherein microdomains of oil and water are interdispersed within the system

In all three types of nanoemulsions, the interface is stabilized by an appropriate combination of surfactants and/or cosurfactants. The emulsions exhibit excellent kinetic stability and are thermodynamically unstable and eventually phase-separate. Emulsions are cloudy, while nanoemulsions are clear or translucent.

39.3 Advantages of a Nanoemulsion over Other Dosage Forms

- It increases the rate of absorption and eliminates variability in absorption.
- It helps in solubilizing a lipophilic drug.
- It is used in an aqueous dosage form for water-insoluble drugs.
- It increases the bioavailability of the drug.

- Various routes, like tropical, oral, and intravenous, can be used to deliver the product.
- It allows rapid and efficient penetration of the drug moiety.
- It is helpful in taste masking.
- It offers protection from hydrolysis and oxidation as the drug-in-oil phase in the oil-in-water nanoemulsion is not exposed to attack by water and air.
- Liquid dosage increases patient compliance.
- Less amount of energy is required.
- A nanoemulsion is a thermodynamically stable system where its stability allows self-emulsification of a system with unique properties.
- A nanoemulsion holds both lipophilic and hydrophilic drugs.
- A nanoemulsion as a delivery system improves the efficacy of a drug, allowing the total dose to be reduced and thus minimizing side effects.

39.4 Disadvantages of Nanoemulsion-Based Systems

- A large concentration of the surfactant and cosurfactant stabilizes the nanodroplets.
- This system offers limited solubilizing capacity for high-melting substances.
- Due to higher kinetic energy, nanoparticles of nontoxic surfactant could cause higher shear in pharmaceutical applications.
- Nanoemulsion stability is influenced by environmental parameters such as temperature and pH. These parameters change upon nanoemulsion delivery to patients.

Thus chemistry for a sustainable future has been the real need and a functional interface of fundamentals of chemistry and its expression in learning, industrial applications, and developing new models in areas of nanosciences, especially nanoemulsion. The chemical reactions and interactions which are conducted and materialized by optimizing the working molecular coordinates (WMC) like functional groups, hydrophobicity, hydrophilicity, temperature, pH and others are significant. Therefore an adequate optimization of WMC is the most challenging task in chemical processes. Their accurate tracking and quantification further add a new feather to the learning and implementation of modern chemical sciences. Accordingly, the friccohesity of molecular activities versus compatibility of tracking tools has been a practical approach of the chemical sciences. To complete the same a few glimpses of the physicochemical potential of chemical systems will be discussed. Presently, the theme of friccohesity as a trump card of modern sustainable approach for an integrated tracking route for the coordinates has become the fundamental need to evaluate, integrate, and develop the molecular models of nonideal chemistry or the chemistry of secondary bonds. The survismeter versus nonideal chemistry will be elaborated in the lecture in light of the WMC.

Chapter 40

Formation of a Nanoemulsion

A nanoemulsion is a multiphase colloidal dispersion and is characterized by stability and clarity. There is an application of high shear obtained by the microfluid or ultrasonic approach used to reduce the droplet size to the nanoscale. A marginal difference between nanoemulsions and microemulsions exists as a micellar phase or mesophase. Microemulsions are formed through thermodynamic self-assembly, whereas nanoemulsions require an external shear for rupturing the droplets. Such driving forces are generated by the hydrophobic forces of the surfactants that approach the larger-sized droplets of the oil. A microemulsion is structurally between 1 and 100 nm. Microemulsions are not emulsions of microscale droplets as they are formed by a self-assembled equation phase where the surface tension does not play a significant role. Nanoemulsions underline the basic principle in microemulsion formulation. They comprise two immiscible phases with interfacial tension (IFT) between them reduced by the addition of surfactant and cosurfactant along with molecules that compensate the hydrophilic balance needed for developing hydrophilic moieties to surround the hydrophobic circular arrays. The critical micelle concentration (CMC) could be easily monitored to optimize the hydrophilic and hydrophobic balances.

Survismeter: Fundamentals, Concepts, and Applications
Man Singh and Sunita Singh

ISBN 978-981-4774-70-3 (Hardcover), 978-0-429-02761-1 (eBook)
www.panstanford.com

Components of a Nanoemulsion: The main components of nanoemulsions are as follows:

- Oil or fat
- Surfactant and cosurfactant
- Aqueous phase
- Chemical stabilizers

40.1 Factors Affecting Nanoemulsion Formulation

- Appropriate composition is required to avoid Ostwald ripening, and the dispersed phase should be highly insoluble in the dispersed medium.
- The dispersed phase should not develop a solvation sphere except realigning the solvent around it.
- The surfactant as an essential part of the nanoemulsion should not form lyotropic liquid crystalline microemulsion phases.
- Systems containing short chain alkanes, alcohols, water, and surfactants form the phases that are generally used with the cosurfactant.
- The presence of excess surfactants enables the new surface area on the nanoscale to be rapidly coated during emulsification, thereby inhibiting induced coalescence.
- Extreme share is applied to rupture microscale droplets to nanoscale by providing a stress level that is more than the Laplace pressure of the droplets with a pressure of 10–100 atm. Out of various methods ultrasonication is widely used in laboratory.

40.2 Nanoemulsion Preparation

Nanoemulsions have a smaller particle size and are most effectively produced using high-pressure equipment. The most commonly used methods for producing nanoemulsions are high-pressure homogenization and microfluidization, used on both laboratory scale and industrial scale. Methods like ultrasonication and in situ emulsification are also suitable for the preparation of nanoemulsions.

40.3 High-Pressure Homogenization

High-pressure homogenization assists in nanoemulsion preparation using a high-pressure homogenizer and a piston homogenizer to produce nanoemulsions of ~1 nm particle size with $\Delta G < 0$. The dispersion of oily and aqueous phases is achieved by forcing the mixture through a small inlet orifice at 500 to 5000 psi, a high pressure. An intense turbulence and hydraulic shear produces NPs of the used materials, which is indicated with the friccohesity data. The NPs exhibit a liquid, lipophilic core separated from the surrounding aqueous phase by a monomolecular layer of phospholipids. It consumes high energy, and there is an increase in the temperature of the emulsion during processing.

40.4 Variables for Optimized Formulation

40.4.1 Homogenization Pressure

The homogenizing pressure for developing nanoemulsions is important to induce adhesive forces around the chosen chemical entity like magnetic nanoparticles capped with citric acid. Therefore, such pressures optimize the cohesive and adhesive forces in the processes. Thus this pressure is noted as a critical parameter and remains operational within 100 to 150 bars. The higher the size, the lower is the particle size (~22 nm). The higher the homogenization cycles, the smaller is the particle size, so the cycles are carried out in sets of 3, 4, or 10 cycles. A polydispersity index depicts the number of cycles analyzed for the drug after each cycle.

Advantages

- Ease of scale-up and little batch-to-batch variation with respect to the nature of medium, additives like glycerol, coadditives like ethanol, pH, temperature, and pressure.

- Narrow size distribution of the nanosize drug
- Flexibility in handling the drug quality
- Effectively used for thermolabile substances

40.4.2 Microfluidization

Microfluidization is a mixing technique that makes use of a device called a microfluidizer. This device uses a high-pressure positive displacement pump (500 to 20,000 psi) that forces the product through the interaction chamber.

For microfluidization microchannels are used where the product flows through them on to an impingement area, resulting in very fine particles (in the submicron range).

40.5 Microfluidizer for Nanoemulsions

The aqueous (hydrophilic) and oily (hydrophobic) phases of two separate solutions join together and are processed in an online homogenizer to form a coarse emulsion. The emulsion is passed into a microfluidizer for further processing for getting stable nanoemulsions (Fig. 40.1). The emulsion is passed via an interaction chamber microfluidizer repeatedly to get particles of the desired size. The bulk emulsion is then filtered through a filter under a nitrogen environment to remove large droplets, resulting in a uniform nanoemulsion. The survismeter renders valuable services as it tracks the particle size. The higher viscosity indicates nanoparticles

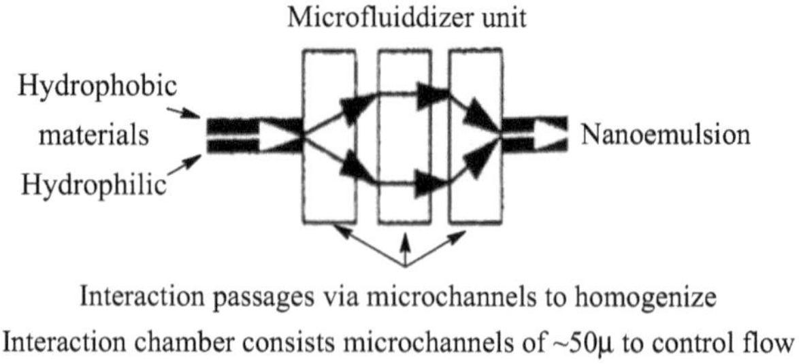

Figure 40.1 Process for getting stable nanoemulsions.

(NPs). It is better to determine the friccohesity because you do not need density for its calculation. An increase in friccohesity indicates NPs.

40.5.1 Ultrasonication

Nanoemulsion preparation uses ultrasonic sound frequency for the reduction of droplet size. The use of a constant-amplitude sonotrode at system pressure in excess of the ambient value reduces the droplet size. The reduction in size weakens cohesive forces (CFs) where the friccohesity becomes exceptionally higher. An increase in external pressure increases the cavitation threshold within an ultrasonic field. It is depicted by the formation of fewer bubbles. An increase in external pressure increases the collapse pressure of cavitation bubbles, and it becomes stronger and more violent in atmospheric conditions. The cavitation is the most important mechanism of power dissipation in a low-frequency ultrasonic system. Such changes in navigational intensity could be related directly to changes in power density.

Phase inversion: A surfactant or a mechanical method is used to weaken the CFs or the chemical energy to initiate NP dispersion for phase transition for emulsification. On varying the composition of the emulsion at a constant temperature or varying the temperature keeping the emulsion composition constant, a phase transition is produced. Several researchers in areas of nanoemulsions have studied the phase inversion temperature where an increase in

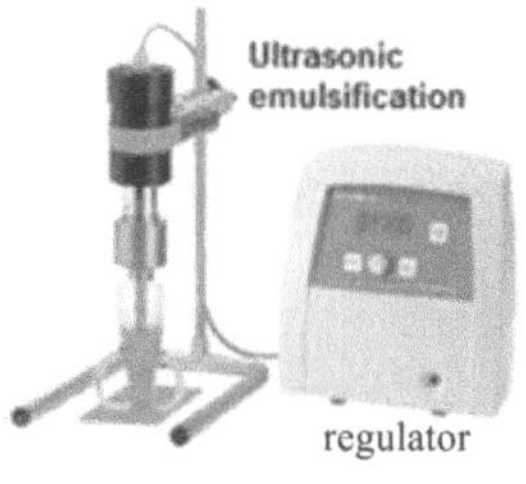

Figure 40.2 Ultrasonication of hydrophobic and hydrophilic liquid mixtures mechanochemical dispersion regulating sonicating frequency as per need for monodispersion of molecules.

temperature results in chemical changes of polyoxyethylene by degradation of the polymer chain with temperature.

40.6 Spontaneous Emulsification

It is performed in three major steps, noted as under:

(1) Prepare a homogeneous organic dispersion of oil and a lipophilic surfactant in a water-miscible solvent and a hydrophilic surfactant.
(2) Magnetically stir the organic phase (oil) in the aqueous phase (water) to form an emulsion.
(3) Evaporate the water from the water-miscible solvent under reduced pressure.

40.6.1 Solvent Evaporation

It involves preparing a solution of the drug followed by its emulsification in another liquid, a liquid that is nonsolvent for the drug. The solvent evaporation precipitates the drug. Crystal growth and particle aggregation are controlled by creating high shear forces using a high-speed stirrer. Thereby, the friccohesity is measured at a specific time interval to depict the shear forces.

40.6.2 Hydrogel Method and Transmembrane Permeation

Hydrogen is obtained on solvent evaporation where the drug solvent is miscible with the drug antisolvent. The higher shear force prevents crystal growth and Ostwald ripening.

Nanoemulsion permeation involves various factors, like:

- Solubility of drug
- Hair follicles in percutaneous absorption
- Presence of stratum corneum and a lipid layer

The surfactant and cosurfactants play a key role in nanoemulsion formation.

Chapter 41

Characteristics and Applications of Nanoemulsions

Droplet size, surface tension, viscosity, density, turbidity, refractive index, friccohesity, phase separation, and pH measurements characterize a nanoemulsion. The droplet size distribution of nanoemulsion vesicles is measured by a light scattering technique, electron microscopy, or the survismeter [8, 9].

41.1 Dye Solubilization and Dilutability Test

A water-soluble dye is solubilized within the aqueous phase of a w/o globule contrary to dispersion in an o/w globule. An oil-soluble dye is solubilized within the oil phase of the o/w globule but is dispersible in w/o globule [10–12].

The o/w nanoemulsions are dilutable with water whereas w/o are not and undergo phase inversion into o/w nanoemulsions [8, 13, 14].

41.2 Conductance in Nanoemulsions

Contrary to w/o the o/w nanoemulsion is highly conducting, where the placement of water matters a lot. In w/o the water is an

Survismeter: Fundamentals, Concepts, and Applications
Man Singh and Sunita Singh

ISBN 978-981-4774-70-3 (Hardcover), 978-0-429-02761-1 (eBook)
www.panstanford.com

internal or dispersal phase. For determining the nature of the continuous phase and to detect phase inversion phenomena, the electrical conductivity measurements are useful [10, 15]. A sharp increase in conductivity in certain w/o nanoemulsions is seen at low volume fractions, and such behavior is interpreted as an indication of percolative behavior or exchange of ions between droplets before the formation of bicontinuous structures. The dielectric measurements depict both structural and dynamic features of nanoemulsions [16, 17].

41.2.1 Dynamic Light Scattering

These measurements are taken at 90° in a dynamic light scattering (DLS) spectrophotometer, which uses a neon laser of wavelength 632 nm. The data processing is done in a built-in computer with the instrument [18, 19].

41.2.2 Polydispersity

The average diameters and polydispersity index of samples are measured by photon correlation spectroscopy. The measurements are made at 25°C using a He-Ne laser [10, 20].

41.2.3 Phase analysis

The identification of nanoemulsions (o/w or w/o) is made by measuring electrical conductivity using a conductometer [15, 21].

41.2.4 Interfacial tension

The properties of a nanoemulsion are studied by measuring the interfacial tension (IFT) using the survismeter. Ultralow values of IFT indicate mutual dispersion of oil and water phases where the surfactant develops hydrophilic and hydrophobic interactions. So the surfactant could be placed at middle-phase nanoemulsions in equilibrium with aqueous and oil phases [8].

The survismeter measures ultralow IFT values. IFT values are derived from the measurement of the number of drops of high-density liquid (HDL) in a low-density liquid (LDL) [13].

41.2.5 Viscosity Measurement

The viscosity of several compositions is measured at several shear rates at variable temperatures using the survismeter. The sample in the survismeter is maintained at a chosen temperature, like 37°C ± 0.2°C by a thermobath. The samples for measurement are taken in the survismeter reservoir bulb (RB). Their pH is measured by a pH meter.

41.2.6 In Vitro Skin Permeation

Skin is placed between the donor and the receiver chambers of vertical diffusion cells. The receiver chambers are filled with fresh water containing 20% ethanol. The receiver chambers are set at 37°C, and the solution in the receiver chambers is stirred at 280–320 rpm. The formulations are placed in the donor chamber. At 2 h, 4 h, 6 h, and 8 h, 0.5 mL of the solution in the receiver chamber is removed for gas chromatography (GC) analysis and replaced immediately with an equal volume of fresh solution. Each sample is performed three times [9]. Cumulative corrections are made to obtain the total amounts of drugs permeated at each time interval. The cumulative amounts of drug permeated through any skins are plotted as a function of time. The permeation rates of the drug at a steady state through live skins are calculated from the slope of the linear portion of the cumulative amount permeated through the live skins per unit area versus time plot. Prior characterization for better efficiency is made by friccohesity data.

41.2.7 Thermodynamic Stability

For thermodynamic stability of drug-loaded nanoemulsions a few stress tests are needed as mentioned below.

***Heating cooling cycle*:** Nanoemulsion formulations are exposed to six cycles between refrigerator temperature (4°C) and 45°C. Later the stable nanoemulsion formulations are subjected to a centrifugation test at 3500 rpm, and those that do not show any phase separation are taken for a freeze thaw stress test.

Freeze thaw cycle: The formulations are subjected to three freeze thaw cycles between 21°C and +25°C under standard laboratory conditions. These studies are made in a 3-month period at 10-day intervals [8, 10].

41.3 Stability of Nanoemulsions

The stability largely depends on the following:

(1) Coalescence of two droplets due to surface tension and intermolecular attractions, which is reduced with surfactants.
(2) High solubility of dispersed phase catalysis diffusive migration of smaller droplets with low Laplace pressure to larger droplets of high Laplace pressure, known as Ostwald ripening. The oil as dispersed phase is selected with a continuous phase for Ostwald ripening. The instability of the nanoemulsion accounts for creaming, flocculation, coalescence, and Ostwald ripening. Ostwald ripening is the main mechanism of nanoemulsion instability because others are minimized by the small size of the nanoemulsion particles and the use of nonionic surfactants. Creaming of the nanoemulsion is prevented by a faster diffusion rate of smaller droplets. Van der Waals force causes attraction of droplets for flocculation as it depicts stronger cohesion with higher surface tension and lower viscosity. The nanoemulsion of a nonionic surfactant does not create any attractive force, so no flocculation occurs.

41.4 Applications of Nanoemulsions

41.4.1 Parenteral Delivery

Nanoemulsions have advantages where intravenous administration is concerned due to a strict requirement of this route of administration of a droplet size lower than 1 μm. The parenteral (injectable) administration of a nanoemulsion is used for a variety of purposes, for example, to inject fats, carbohydrates, and vitamins and other nutrition [8, 21]. Nanoemulsions of soybean, sesame, and olive natural oils with nontoxic surfactant pluronic F-68 via ultrasound are useful for parenteral feeding. Lipid nanoemulsions have been widely explored for parenteral delivery of drugs. Nanoemulsion formulations have distinct advantages over microemulsions when delivered parenterally because a nanoemulsion is cleared more slowly than the coarse particles of an emulsion and has a longer residence time in the body [22].

41.4.2 Oral Delivery

Nanoemulsion formulations offer several benefits over conventional oral formulation for oral administration, including increased absorption, improved clinical potency, and decreased drug toxicity. Therefore, a nanoemulsion is ideal for delivery of drugs such as steroids, hormones, diuretics, and antibiotics. Pharmaceutical drugs of peptides and proteins are highly potent and specific in their physiological functions [23].

Primaquine when incorporated into an oral lipid nanoemulsion reportedly showed effective antimalarial activity against *Plasmodium berghei* infection in mice at a 25% lower dose level as compared to conventional oral dose. A lipid nanoemulsion of primaquine improved oral bioavailability in the liver with a drug concentration higher by at least 45% as compared to the plain drug [24].

41.4.3 Topical Delivery

A direct and targeted drug delivery to the affected area of the skin or eyes is needed where a nanoemulsion could achieve a level of topical antimicrobial activity achieved by systemic antibiotics. The nanoemulsions show a wider spectrum of activity against bacteria (*Escherichia coli* and *Staphylococcus aureus*) and fungi (*Candida* and dermatophytes) [15, 25, 26].

41.4.4 Cosmetics

The esthetic properties because of a low viscosity and transparent visual aspects of a nanoemulsion with droplet sizes <200 nm seem functional. High surface areas of nanoemulsions for effective transport of active ingredients to the skin make nanoemulsions significant for application in cosmetics [27]. Stable nanoemulsions are useful in cosmetics because they do not cause inherent creaming, sedimentation, flocculation, or coalescence. The incorporation of potentially irritating surfactants is avoided by using high-energy equipment during manufacturing. Nanogel technology to create miniemulsions from oil-in-water concentrates minimizes transepidermal water loss and enhances skin protection and penetration of active ingredient. It is useful for sun care products, moisturizing, and antiaging creams [28].

41.4.5 Transdermals

Generally, drug delivery in nanoemulsions is safer due to controlled philicphobic thin film around drug molecule. In the light of the above, indomethacin, a potent nonsteroidal anti-inflammatory drug (NSAID), its anti-inflammatory effects in its nanoemulsions is compared with the marketed gel in carrageenan-induced paw edema in rats. The percentage inhibition value that the nanoemulsion develops has a great potential for transdermal application of indomethacin. A nanoemulsion that contains 2% celecoxib, 10% oil phase (sefsol 218 and triacetin), 50% surfactant mixture (Tween 80 and Transcutol P), and 40% water seems attractive for transdermal delivery of celecoxib. An anti-inflammatory effect and percentage inhibition value 24 h after administration is high for nanoemulsions (81.2%) as compared to celecoxib gel (43.7%) and nanoemulsion gel (64.5%). In vitro and in vivo studies show a significant increase in anti-inflammatory effects of aceclofenac nanoemulsion (82.2%) as compared to nanoemulsion gel formulation (71.4%) and conventional gel (41.8%) [29, 30].

41.4.6 Biotechnology

The enzymatic and biocatalytic reactions are conducted in pure organic or aquaorganic media, where biphasic media is used for such reactions. The use of pure, apolar media causes the denaturation of biocatalysts. The use of water-proof media is relatively advantageous [31]. Enzymes low in water content display the following activities.

- Increased solubility in nonpolar reactants
- Shifting thermodynamic equilibria in favor of condensations
- Improved thermal stability of enzymes, enabling reactions at higher temperatures [32, 33]

41.4.7 Determining van der Waals Forces in DMSO, Water, and DMF

The stronger the surface tension, the higher are the van der Waals forces among similar molecules; also the lower the surface tension, the weaker and higher is the viscosity and the weaker are the cohesive forces (CFs) or the van der Waals forces. Such experiments could be conducted with several salt and amino acid aqueous liquid mixtures.

Such an approach could be extended for the droplet size of a nanoemulsion that prevents flocculation as small droplets show a high curvature and Laplace pressure opposes the deformation of large droplets. Coalescence of droplets of a nanoemulsion is prevented by a thick multilamellar surfactant film adsorbed over an interface of the droplets. Ostwald ripening causes instability of the nanoemulsion. In Ostwald ripening small droplets with a high radius of curvature convert into large droplets with a low radius of curvature. Two droplets diffuse and become one large droplet. Thus on storage for a long time period, the droplet size distribution shifts to large sizes and the transparent nanoemulsion becomes turbid. Hence Ostwald ripening creates trouble during the delivery of formulations. A polymeric surfactant prevents Ostwald ripening at the interface, which increases the elasticity of droplets and further reduces the effect of Ostwald ripening [9]. Nanoemulsion formulations offer several advantages for the delivery of drugs and as biological or diagnostic agents and are being used in clinics for long as these are total parenteral nutrition fluids. Diprivan, Liple, and Ropion drug delivery has reached the marketplace. NEs are seen as vehicles for administering aqueous insoluble drugs as they have recently received increasing attention as colloidal carriers for targeted delivery of various anticancer drugs, photosensitizers neutron capture therapy agents, and diagnostic agents. Because of their submicron size, they can be easily targeted to the tumor area. Several research papers have been published for improving drug delivery.

Chapter 42

Friccohesity and Its Applications

Friccohesity distinguishes the electrostatic forces that induce interacting abilities. For example, proteins have stronger cohesive forces (CFs) and water weakens them because of water–peptide bond formation with different electron clouds, which influence the friccohesity. The CF is a force that operates within similar electron clouds on the bond Schrödinger boundary or similar boundaries, but intermolecular forces (IMFs) work with different electron clouds and align along the different boundaries of bonds within dissimilar molecular forces, as in the aqueous dimethylformamide (DMF) solution. Similarly, the rhodamine b dye solutions in water and aqueous acetic acid develop IMFs with a definite friccohesity.

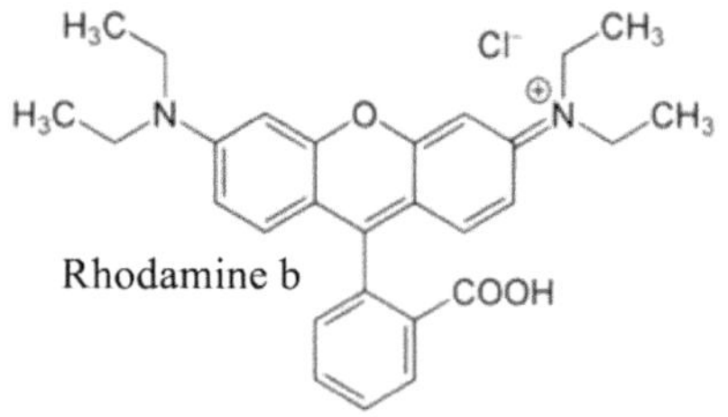

Rhodamine b

Figure 42.1 Fluorescent rhodamine b dye efficiently interact with web-drimers (dendrimers) and develop functional liquid mixtures.

Survismeter: Fundamentals, Concepts, and Applications
Man Singh and Sunita Singh

ISBN 978-981-4774-70-3 (Hardcover), 978-0-429-02761-1 (eBook)
www.panstanford.com

42.1 Determination of the Nanoemulsion of $BaSO_4$

Contrary to $BaSO_4$, barium sulfide (BaS) is soluble in water and is readily converted to oxide, carbonate, and halides. To produce pure barium sulfate, the sulfide or chloride is treated with sulfuric acid or sulfate salts.

$$BaS + H_2SO_4 \rightarrow BaSO_4 + H_2S$$

$BaSO_4$ is produced as permanent white and is noted as Blanc fixe, which is a form of barium encountered in consumer products like paints. $BaSO_4$ is obtained on combining solutions of barium ions and sulfate salts. $BaSO_4$ is less toxic due to its insolubility. Wastes containing barium salts are treated with sodium sulfate to immobilize (detoxify) the barium. Its low solubility is exploited in qualitative inorganic analysis for testing Ba^{2+} ions as well as for sulfate. Baryta, a thin layer of $BaSO_4$, is first coated on the base surface of photographic paper in most cases to increase the reflectiveness of the image, with the first such paper introduced in 1884 in Germany. The light-sensitive silver halide emulsion is then coated over the baryta layer. The baryta coating limits penetration of the emulsion into fibers of paper and makes the emulsion more even, resulting in more uniform blacks. Further coatings are for fixing and protection of image. Baryta is used to brighten papers intended for inkjet printing, so nanoemulsions of $BaSO_4$ could be developed for films of a better quality. A nanoemulsion could be developed by using NH_4OH, PEG, or water and less amount of $BaSO_4$. The particles' size and surface area could be measured using the survismeter.

42.2 Questionnaires

There are two peculiar situations:

The salts/ions solubilize in water through structure breaking, but structured solvents surround the saturated alkyl chain on dispersion. So oil is not dissolved in water, but the surfactant activates the oil so it disperses in water. However, the length of

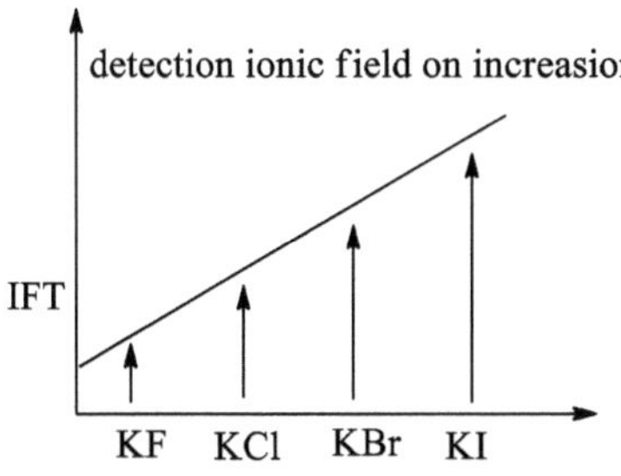

Figure 42.2 IFT vs size of anions with the common cation (K^+) infers a role of anion from F^- to I^- on mutual solubilization of water and organic solvents like benzene, toluene, ethylbenzene, and hexane. The increasing IFT with increasing size of anions from F^- to I^- infers their weaker mutual solubilization because of their stronger interactions with water especially induced dipoles interactions of the I^- that develop stronger interactions with the water. It is the best model to determine quantitatively the induced dipoles interactions and distinguish them from van der Waals forces driven interactions.

the alkyl chain of fatty acids in oil and that of the surfactant affect dispersion dynamics. The survismeter determines the carbon numbers in the surfactant using the following equation:

$$\sigma = \sigma^0 \pm n(-CH_2-)$$

Here, σ is the friccohesity, σ^0 is the limiting friccohesity, and n is the number of carbon atoms or the number of $-CH_2-$ groups in the alkyl chain. Initially, the survismeter is calibrated with the surfactants with an increasing alkyl chain. The survismeter determines the hydrophilic group of surfactants, like quaternary ammonium nitrogen (QAS), sodium dodecyl sulfonate (SDS), philic-phobic (hydrophilic and hydrophobic), and nonionic, using friccohesity. Friccohesity determines the degree of solubilization, conjugation, coagulation, and coalescence of the nanoemulsion as the solvent weakens its CF and establishes an IMF. Similarly, the survismeter can detect van der Waals forces on increasing the anionic sizes (Fig. 42.2).

The increasing ionic sizes engage larger solvent molecules, which do not permit molecules to take part in mutual solubilization. Thereby, a few salient features are associated with the survismeter, noted as under:

- It is a device that produces dynamic and functional data.

- It enables the production of live and interconnecting data.
- It can be used for producing industrial and molecular activity tracking data.
- It helps produce nanoparticle (NP) characterizing data.
- It produces material formulating data and food texture determining data.
- It helps identify NP capping.
- It produces university industry interacting data.
- It produces data in day-to-day use.
- It formulates industrial and thermodynamic applications of nanoemulsions.
- It synchronizes the structural geometry of a liquid material matching structural geometry of survismeter.
- It enables minimized used of pesticides for crops and has medical applications (e.g., testing of jaundice).
- It estimates surface tension and viscousness of lung fluid.

Question: What are the factors that reduce the resolution of measured data? How such situation does make surface energy or tension highly resolved data?

Chapter 43

Novelty in Polarity of Solutions and Nanoemulsions

The entire device that creates a hurdle in the movement of ions or molecules within a continuous liquid phase contrary to the surface energy faces the extra natural resistance of the medium networking. So the conductance, pH, density, potential, dynamic light scattering (DLS), high-powered liquid chromatography (HPLC), and others fall in this category, which constitute a continuum model of liquid mixtures during operations processes. However, the surface energy measures depend on the noncontinuum model of liquid mixtures, where the resistance of liquid is cut off, especially in the case of the survismeter. The pdn formation cuts off the relation with continuum liquid flow anchored on the continuum rigid glass boundary. The pdn is formed when there is the least resistance caused by the medium, that is, air, so experimental restrictions are automatically cut off or deleted in measuring the surface tension. So the surface tension data are the most resolved and accurate, provided the surface tension is measured cautiously. However, such precautions are already taken care of in the case of the survismeter device. The DLS could face less resistance due to a nonstructured liquid mixture and Brownian motions. The viscosity is convection based

Survismeter: Fundamentals, Concepts, and Applications
Man Singh and Sunita Singh

ISBN 978-981-4774-70-3 (Hardcover), 978-0-429-02761-1 (eBook)
www.panstanford.com

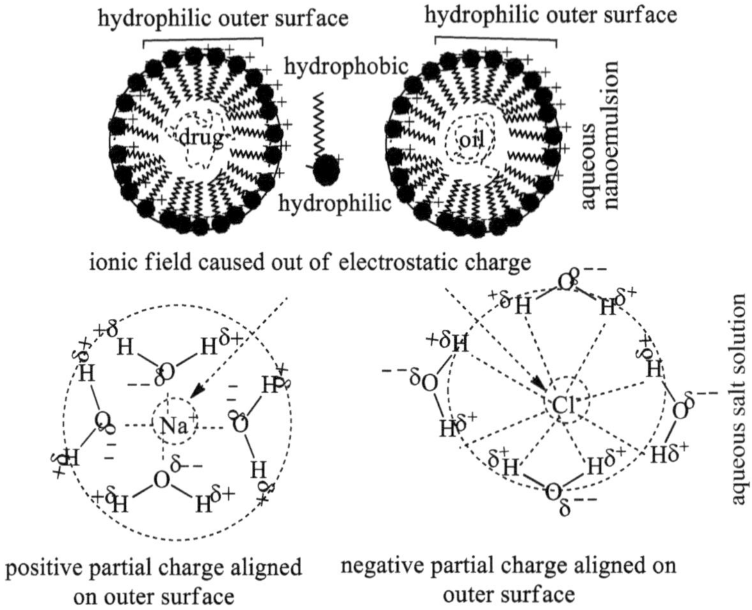

Figure 43.1 Structural dynamics of micelle formation that optimizes the hydrophilic and hydrophobic alignments and interactions with certain structural potential. Normal micelles with a hydrophobic environment of core align and traps drug into its cores while the oil particles also move towards core due to an adequate hydrophobicity. The effect of NaCl salt is seen to affect the critical micellar concentrations (CMC).

on mass transfer. The mass in parts in connectivity is transferred from one to another point, so it could have less restriction except the friction from a rigid wall. In the year 2001 the survismeter was shaped up, which depicts and distinguishes the aqueous salts from nanoemulsions. The salt solutions induce stronger coulombic forces on realigning the solvent charges by ions, but hydrophilic charges exist on the outer surface of the normal micelle scale.

43.1 Thermodynamics of Nanoemulsions

Electrolytes realign negative and positive charges of water, which enhance cohesive forces (CFs) with a lower friccohesity. Such arrangements develop ionic hydration spheres with certain elements of the ionic field. In such an environment the surfactant accommodating capacity of the solvent is decreased so micelle formations are attained at lower surfactant concentrations. Since the ionic strength of salt works well in an aqueous solution, when a surfactant is added a stronger ionic field further engages more water. So micelles are formed at a slightly lower surfactant concentration. Here, Debye–Hückel theory works very well in the case of micelle formation. It could also be connected with the normal micelle or the reverse micelle formation. In such cases the CFs and distributing forces of the surfactant are structurally influenced. Their product is noted as friccohesity. Friccohesity gives solutions of critical micelle concentration (CMC) relationship with the ionic field of ion hydration in an aqueous medium. Friccohesity is experimentally determined using the survismeter.

Experiments

(1) Determine oxygen-capturing ability (OCA) of nanoemulsions of a series of cationic, anionic, and nonionic surfactants using the survismeter.

Theory: A NE captures air, and the O_2 could be passed into the reservoir bulb (RB) of the survismeter. The reading could be repeated several times till they produce a constant pdn.

(2) Determine the surface tension of the hypophase (protein) + phospholipid (dipalmitoylphosphatidylcholine) + calcium ions and also the OCA of the resultant liquid mixture.

Chapter 44

CO_2 Determination of a Healthy Person Using the Survismeter

A person exhales during breathing certain gases that could be passed to a surfactant solution through a silicon tube for measuring the friccohesity. Several doses could be exhaled into the same solution for further measuring the friccohesity. A gas doses versus friccohesity plot could predict the impact of the gases.

Experiments

(1) Determine the ability of alkanethiol to cap Au nanoparticles (NPs) while increasing the alkanethiol's chain length.
(2) Determine the capping capacity of polyvinylpyrrolidone (PVP) while increasing its molecular weight, using the survismeter.
(3) Determine the activation energy of PVP on increasing its molecular weight.
(4) Determine the activation energy of monobasic (CH_3COOH), dibasic (malonic acid), and tribasic acid (citric acid) in aqueous solutions.
(5) Determine the activation energy of H_3PO_4, NaH_2PO_4, and Na_3PO_4 in aqueous solutions.

Survismeter: Fundamentals, Concepts, and Applications
Man Singh and Sunita Singh

ISBN 978-981-4774-70-3 (Hardcover), 978-0-429-02761-1 (eBook)
www.panstanford.com

(6) Determine van der Waals forces of a series of amino acids, carbohydrate, alcohol, carboxylic acids, amines, proteins, and others using the survismeter.
(7) Determine the effect of mono- and polydisperse liquid samples in the body. [*Hint:* Monodispersed food is naturopathic in nature but polydispersed food like starch initially breaks down into its simple form.]

Electrolytes realign negative and positive charges of water, which enhances cohesive forces (CFs) with lower friccohesity. Such arrangements develop ionic hydration spheres with certain elements of an ionic field. In such an environment the surfactant-accommodating capacity is decreased so micelles are formed at lower surfactant concentrations. The ionic strength of salt could develop a relation with the hydrophilic activity of a surfactant engaging more water. So when the the size of the cation increases with an increasing size of the alkyl chain of the surfactant, it could have some effect on micelle formation. Such approaches assist in micelle formation at a lower surfactant concentration. These effects are quantitatively determined with friccohesity measurements. The strength of hydrogen bonds among different molecules is determined by friccohesity because the friccohesity drastically increases on stronger hydrogen bonding (HB). Friccohesity gives solutions of ionic field disturbed by ion hydration in an aqueous medium.

Questions

(1) Is there any relationship between critical micelle concentration (CMC) and Debye length?
(2) How do we understand multiphase wettability from a molecular point of view?
(3) How can we develop nanoemulsions that will remove toxins from the body as they dissolve out?
(4) How can we use the survismeter to determine the size of the cations and anions?
(5) How can we use the survismeter to determine the length of the alkyl chain in a homologous series of alcohols?

(6) How can we use the survismeter to determine the length of the alkyl chain in surfactants?
(7) How can we use the survismeter to determine the degree of HB?
(8) How can we distinguish van der Waals forces from London dispersive forces (LDFs) using the survismeter?
(9) Do dendrimers act as natural micelles as they have core and peripheral structures? [*Hint:* Micelles hold the drug at the core, as do dendrimers.]

The artificial situation of Couette flow is a fluid flowing via a rigid and stationary capillary. The fluid right next to the capillary's rigid wall is at rest, and the fluid at the center of the capillary moves faster than that nearer the wall. Let q be the volume flow rate ($q = Av$). So the volume flow rate of a viscous fluid is given by:

$$q = \pi R4\Delta P8\eta L \text{ (Hagen–Poiseuille equation)}$$

In the years 1839 and 1838, German physicist and hydraulic engineer Gottlif Hagen and French physician Jean Marie Louis Poiseuille developed the equation, respectively. Here, R is the radius and $L$is the length, with a pressure difference of ΔP between the ends of the capillary. The viscosity is denoted by the Greek letter eta (η).

44.1 Lungs and Friccohesity

Lungs are the primary organs of respiration in humans, where the two lungs (the right lung and the left lung) are located near the backbone on either side of the heart. Their function in respiration is to extract O_2 from the atmosphere and transfer it into the bloodstream and to release CO_2 from the bloodstream into the atmosphere in a process of gas exchange. In humans, the diaphragm is a primary muscle that drives breathing. The lungs provide the airflow for making vocal sounds and human speech. The lungs are situated within the thoracic cavity of the chest. The right lung is bigger than the left as the left lung shares space in the chest with the heart. The lungs together weigh ~1.3 kg, and the right is heavier. The lungs together contain ~2400 km of airways and 300 to 500 million alveoli. Each lung is enclosed within a pleural sac, which allows the inner and outer walls to slide over each other without much friction when breathing occurs. The sac, which encloses each lung, divides the lung into lobes. The right lung has three lobes, and the left has two. The lobes are divided into bronchopulmonary segments and lobules.

The lungs have a unique blood supply, receiving deoxygenated blood sent from the heart for the purpose of receiving oxygen (pulmonary circulation) and a separate supply of oxygenated blood (bronchial circulation). The surface tension of the lungs' fluid remains $17 \approx 22$ mN/m so that there is no restriction of CFs on their adequate expansion. The loading of carbon on the lungs' alveoli through smoke or by adding any surface tension–increasing substance in the lungs' fluid due to cigarette smoking, drug addiction, wrong food habits, or inhaling polluted air is always avoided. A number of diseases, like pneumonia and lung cancer, affect the lung tissue. Diseases like chronic obstructive pulmonary disease and emphysema are caused by smoking or exposure to harmful substances. In embryonic development, lungs begin to develop as an outpouching of the foregut, similar to the tube for forming the upper part of the digestive system. When the lungs are formed, the foetus is still held in the fluid-filled amniotic sac. So they do not function to breathe. Blood is also diverted from lungs through a ductus arteriosus. At birth air begins to pass through the lungs and the diversionary duct closes, which helps the lungs to begin to respire.

44.2 Cellulose Acetate Propionate

Experiment: Determine the coulombic forces between two separate molecules developed when HB is developed between them.

The HB between similar molecules, like cellulose acetate propionate (CAP) or cellulose acetate butyrate (CAP), develops CFs as the electron clouds around the molecules remain the same. When two molecules of water develop HB, then their electron clouds remain the same, but when HB is developed between a water molecule and an ethanol molecule, then their electron clouds differ from each other. In such a situation the HB is affected by the electron clouds. A molecular effect of this type could be noted as HB between dissimilar molecules (Hidmol) with definite interacting molecular forces (Imolf) working between them. Electrostatic forces operative in dissimilar molecules ((Elfodim) in the case distinguish the interacting activities of the individual molecules in the mixtures. The applications of such activities could affect the nanodispersions of biofertilizer or compost that are able to enter the root membrane though absorption from the soil. Ultimately a fertilizer has constituents that assist plant growth after the fertilizer's intake from the soil through the root membrane. So nanosize constituents could enter the root member. It is similar to the case where heterohesive forces between extracellular fluid and intracellular fluid boost the intake of desired nutrients through the root membrane.

Few Innovative Experiments

(1) Determine the reduction of H and O atoms from glucose or sucrose either on heating or on using dehydrating agents.

(2) Determine the friccohesity of ethylene-vinyl acetate (EVA) (biomedical engineering application as a drug delivery device) by dissolving dichloromethane.

The effect of sugar on drug binding is a noticeable academic activity because the sugar in liquid solution gets rapidly mixed to form a homogeneous state. Therefore, such studies seem useful, which could lead to a slow release of the drug via CFs, adhesive (heterohesive) forces, and friccohesity. For such systems the density could be determined using the following equations:

$$\rho \propto \left(\frac{dv}{dt}\right)_{\mathrm{T,P,c}} \quad \text{or} \quad \rho = \mathrm{k}\left(\frac{dv}{dt}\right) \tag{44.1}$$

$$\rho = \left[\left(\frac{\sigma}{\sigma_0}\right)\left(\frac{n}{n_0}\right)\left(\frac{t}{t_0}\right)\right]\rho_0 \tag{44.2}$$

Here, ρ and ρ_0 are densities of sample and solvent, respectively; n and n_0 are pdn for sample and solvent, respectively; t and t_0 are viscous flow times for sample and solvent, respectively;

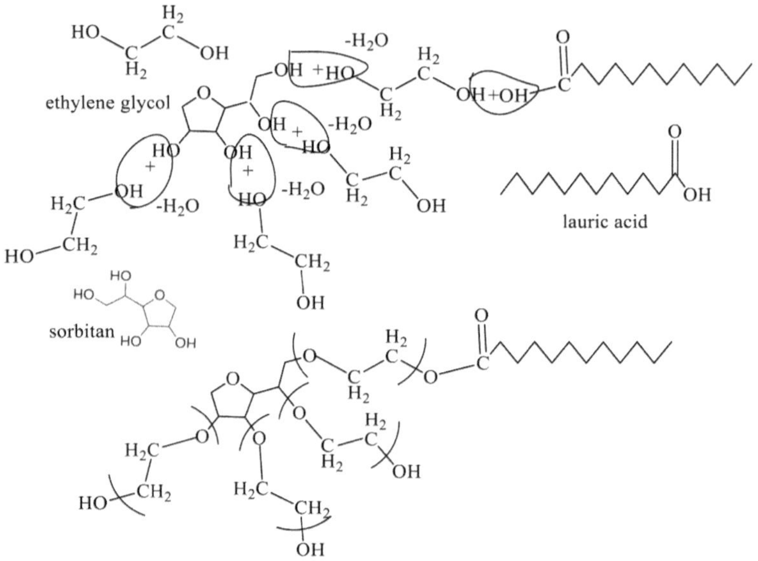

Figure 44.1 The condensation reaction of sorbitan with ethylene glycol to prepare tweens, the nonionic surfactants with adequate numbers of free alcoholic OH groups and alkyl chain.

and σ and σ_0 are the friccohesity values for sample and solvent, respectively.

(3) Determine the CMC of nonionic and anionic forms of surfactants in an aqueous medium and distinguish their chemical mechanisms.

In general, the CMCs of tweens (Tw), sodium dodecyl sulfate (SDS), and dodecyl trimethylammonium bromide (DTAB) are DTAB > SDS > Tw.

***Questions*:**

(1) Why does Tween 20 (Tw20) form the CMC at its lower concentration contrary to SDS, which forms the CMC at a higher concentration?

***Answer*:** As per Fig. 44.1 Tw20 contains three –OH groups, which develop HB three times with the water as solvent and engage larger water molecules. Thus a slightly higher population of Tw molecules starts developing CFs, forming micelles in the absence of their activities with water. SDS has one anionic site and counteraction, which develop ionic interactions with water, which could also develop structure breaking interactions with water. So more and more monomer water is available with SDS to engage larger SDS molecules contrary to the Tw20, which is unable to initiate the structure breaking activities.

Since the CMC formation depends on the availability of the solvent, which keeps dispersing the surfactants, when the solvent is occupied the surfactant starts reorganizing in the form of micelles. Also if the solvent is engaged largely by a fewer number of surfactant molecules like Tw20, then the micellar shape starts to form at a comparatively lower concentration. Thereby the manifold –OH groups of Tw20 engage large water molecules, which result in micelle formation at a lower concentration of Tw20 as compared to SDS. So the CMC could be defined as under:

$$\text{Log CMC} = k \cdot n_\text{f} + c_\text{sur}$$

(2) Under what conditions does the CMC develop in the case of surfactants and saturation develop in the case of salts or polar molecules.

(3) How does the stoichiometry act as a deciding factor for CMC formation?

(4) How does the survismeter distinguish the monodisperse CMC from the polydisperse CMC?

Chapter 45

Noncovalent Interaction Engineering

Noncovalent interaction engineering of molecules is materialized through their electronic clouds, electrostatic dipoles, functional groups, atomic size with electronic configuration, molecular size, and their hydrophobic constituents. Therefore, the structural constituents of the molecules contribute a lot for making the best use of the structures of the chemical substances.

Questions:

- How does the molecular geometry affect the energy barrier?
- What is the relation between knowledge of chemical sciences and human values?
- What is an experiment in chemical sciences?
- What is a research hypothesis?
- What is the basic relationship between UV studies and molecular properties?

Friedrich Wohler's experiment in 1828 contradicted vital forces theory and is noted as a turning point in chemistry. It significantly inspired researchers to develop new molecular structures. Similarly, Kekule gave an insight about the chemical structure of benzene. Linus Pauling extended structural understanding of benzene by

Survismeter: Fundamentals, Concepts, and Applications
Man Singh and Sunita Singh

ISBN 978-981-4774-70-3 (Hardcover), 978-0-429-02761-1 (eBook)
www.panstanford.com

developing a quantum mechanical structure of benzene resulting from resonance.

Question: Between an emulsion and a nanoemulsion, which contains higher energy?

45.1 Hydrophobic Engineering and the Survismeter

The approaches of anionic, cationic, and nonionic surfactants along with their counter ions are different from one another along with their surface charges in the case of their micelle formation. The structural changes of their micelles lead to the development of the zeta potential along with the critical friccohesity. As is noted in Fig. 45.1, the adhesion of water molecules at the periphery of micelles makes a thin film of water molecules. Such adherence of water molecules to micellar surfaces affect surface tension. The molecules, when they are reoriented with the centripetal force, could have a specific electron cloud, but when they move to spread on the surface, they go to different spatial reorientations. The electrospun electron cloud, their density, electrostatic forces like stronger adhesive forces (ADF) induce least entropy. It also develops least kinetic energy (KE), which lead to the development of a single droplet that adheres to the solid surface area. Adherence is due to a variable interfacial energy or a surface energy at a close contact where their energy could be exchanged as the AB-AB transitions occur. Figure 45.2 shows interfacial thermodynamics or friccohesity or entropy changes. Because of molecular forces, a variety of molecular geometries are developed with a variety of IFM, and their split and distribution occurs with definite INNFT and tentropy along with friccohesity. These forces are confined to a finite boundary and hence adhesion takes place.

Similar to the lowest surface tension values of stable fine nanoemulsions the surface tension of supercritical fluids (SCFs) is almost zero. As there is no liquid or gas interface SCFs are kinetically active with the least or zero potential energy. Pressure and temperature both affect the binding energy of the fluid with specific

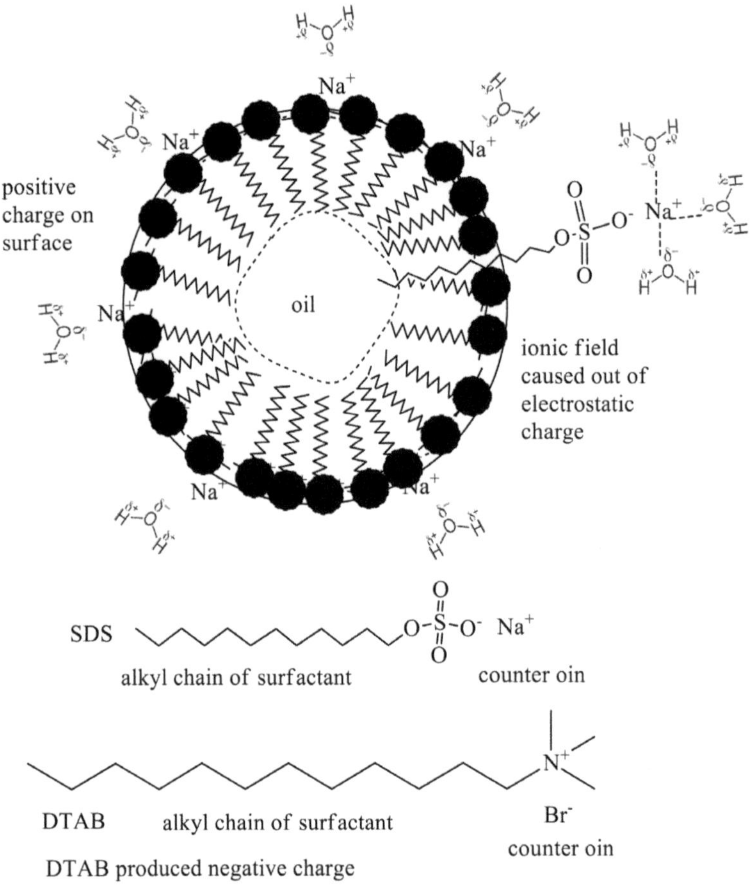

Figure 45.1 Hydrophobic and hydrophilic structural constituents of micelles with their balanced stoichiometry for stability of micelles.

properties that indicate liquid- or gas-like behavior. The solubility of a material in the fluid decides the fate of the SCF. Solubility in a SCF increases with density of the fluid at a constant temperature. The solubility increases when the pressure is increased. However, its relationship with temperature is a multistep process. At a constant density, the solubility increases when the temperature is increased but near the critical temperature point (CTP), the density drops sharply with a slight increase in temperature. Near the CTP, the solubility decreases when the temperature is increased, and then

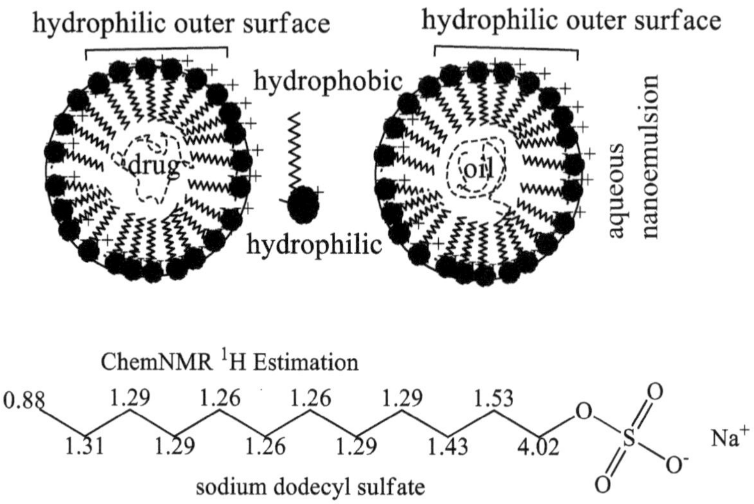

Figure 45.2 H^1NMR infers chemical shielding of H atoms attached to C atoms of alkyl chain of sodium dodecyl sulfate (SDS) that affect micelles formation of oil in water. Sulfate group of SDS affect the hydrophilic as well as hydrophobic interactions of micelles. Similarly the cationic and nonionic surfactants with different hydrophilic functional groups affect the hydrophilic interactions occurring around the outer surfaces of the micelles causing a continuous uniform thin film of hydrogen bonded water through the van der Waals forces.

rises again. SCFs are mutually miscible so for a mixture a single phase could be regained if its CTP is exceeded. The CTP of a binary mixture is estimated as an arithmetic mean of the critical temperature and pressure of two components.

$$T_{\mathrm{c(mix)}} = (x_{\mathrm{A}}) \cdot T_{\mathrm{cA}} + (x_{\mathrm{B}}) \cdot T_{\mathrm{cB}}$$

The CTP is calculated using equations of state like Peng Robinson or group contribution methods. Properties like density are calculated using equations of state. Viscosity, relative permittivity, and solvent strength as physicochemical properties show large gradients with the pressure near the CTP as these are closely related to density. For example, when CO_2 is at 400 K, the density increases almost linearly with pressure, so the survismeter is a useful tool for SCF determination. Friccohesity is introduced by pressure, temperature, mechanical activity, and chemical agents of a resultant liquid mixture. The essential condition of friccohesity is a continuity of the phases. Similar to J. W. Gibbs thermodynamic theory for capillary action the surfaces lead to discontinuity for thermodynamics of 2D surfaces. Two or more than two electronically active chemical species develop friccohesity theory. Gibbs energy (∂G, J/mol) is correlated to the surface tension as under:

$$\partial G = \gamma \partial A \quad \text{or} \quad \gamma = \left(\frac{\partial G}{\partial A}\right)_{\mathrm{T,P}}$$

A higher surface tension shows a lower entropy because of the stronger cohesive forces (CFs) with stronger binding forces. Higher temperature and compositions (molality, m) both affect the surface tension, which is connected to the entropy as follows:

$$-S^A = \left(\frac{\partial \gamma}{\partial T}\right)_{\mathrm{T,P}} \quad \text{and} \quad -S^A = \left(\frac{\partial \gamma}{\partial m}\right)_{\mathrm{T,P}}$$

Most comparatively the salts affect Gibbs energy and the surfactants, especially with longer alkyl chains, affect entropy. There is a close relation between the surface energy and the SCF that is measured with the survismeter. The friccohesity explores this relationship for academic and industrial applications. Thus the higher the entropy the higher is the temperature; also if the reaction is enthalpic then the entropy is more and the surface tension is lower. The survismeter could determine the entropy and enthalpy of the system. The SCFs assist in the extraction of floral fragrance from flowers to be used in food science for creating decaffeinated coffee, functional food ingredients, pharmaceuticals, cosmetics, polymers, powders, bio- and functional materials, nanosystems, natural products, biotechnology, fossil and biofuels, microelectronics, energy, and environment. Recent experimental methods have played an important role in the area of SCFs, with the focus being on the dynamic properties of fluids.

The viscosity of liquid samples directly depends of their density and viscous flow times which is noted in the equation given below.

$$\eta = kt\rho \quad \text{or} \quad k = \frac{\eta}{t\rho}$$

Therefore a format for recording data could be tabulated as under:

m NaCl	Vft	ρ	η	γ
0.05				
0.10				
0.15				
0.20				
0.25				

The liquid flows in the capillary which is explained as:

$$\rho = \rho^0 + at + bt^2$$

Chapter 46

Demulsifier Salts: Pulling out of Impurities by Preferential Binding Activities

Demulsifier salts are a combination of salts that help in determining the concentration of salts in impure water with the help of demulsifier agents added to water. These salts combine with other salts through ionic hydration and settle them, which can be estimated using a suitable technique.

Theory: Demulsifier salts pull out the impurities due to preferential binding activities with the dispersion phase. The salting out kinetics with time are studied with the survismeter. For example, when the dissolved salts are removed from the bulk phase by controlling solubility product of the friccohesity changes. Initially, a calibration curve is developed with known systems, which is extended for unknown samples. Several queries or questions could be raised about the survismeter sciences, a discussion of which will enhance the understanding of the science and operation of the survismeter. A few such questions are noted here.

- How does the survismeter work, and what is its working theory?

Survismeter: Fundamentals, Concepts, and Applications
Man Singh and Sunita Singh

ISBN 978-981-4774-70-3 (Hardcover), 978-0-429-02761-1 (eBook)
www.panstanford.com

- How does the survismeter allow liquid manifolds furcation to several units of the survismeter?
- What is the main concept of the survismeter, which leads to the growth of a whole idea of survismeter science?
- How does a controlled processing unit (CPU), involving vibrant and virtual valve, physical valve (stopper), and fluidity of sample, work?
- How does the survismeter determine the capping of Au nanoparticles (NPs), ZnO, TiO_2, NiO, Ag, magnetic nanoparticles (MNPs)?
- How does the survismeter determine the free radical–scavenging activity of 2,2-diphenyl-1-picrylhydrazyl (DPPH) by curcumin?
- How does the survismeter determine the drug binding and release study?
- How does the survismeter distinguish a solution process from a nanoemulsion process?
- How does the survismeter determine aggregation and coalescence as dynamic processes?
- How does the survismeter illustrate the ionic field of a series of alkalis or alkali salts?
- How does the survismeter determine Gibbs energy?
- How does the survismeter determine micelles and reverse micelles and the extraction of a product from the bulk to micelles or to the bulk phase?

If 10 mL of a solution under a set of experimental conditions is measured with a device and it takes 10 s to flow and 10 mL of another sample within similar conditions takes 1 s, then the viscosity of the second sample is lower because the velocity gradient is faster. Thus the faster the velocity gradient, the lower is the viscosity and the less is the friccohesity.

$$F = \eta A \left(\frac{\partial v}{\partial y}\right) \quad \text{or} \quad \frac{F}{A} = \eta \left(\frac{\partial v}{\partial y}\right) \quad \text{or} \quad \tau = \eta \left(\frac{\partial v}{\partial y}\right)$$

Shear stress depends on the velocity gradient only where the stronger opposing force F develops higher velocity gradient values due to structural changes in liquid mixtures. The symbol η is dynamic viscosity, and the v/y ratio is the rate of shear deformation or shear velocity (SV).

$$F = \eta A \left(\frac{v}{y}\right)$$

This ratio is a derivative of fluid speed in a direction perpendicular to plates. Isaac Newton had expressed viscous forces by the following differential equation:

$$\frac{F}{A} = \eta \left(\frac{\partial v}{\partial y}\right) \quad \text{or} \quad \tau = \eta \left(\frac{\partial v}{\partial y}\right)$$

The shear stress and velocity gradient both are depicted as under:

$$\frac{F}{A} = \tau,\ Nm^{-2} \quad \text{and} \quad \left(\frac{\partial v}{\partial y}\right), s^{-1}$$

Here, $\tau = F/A$ and $\partial v/\partial y$ is the local shear velocity (LSV). The LSV is applicable for a flow moving along parallel lines and the y axis, perpendicular to the flow, points in the direction of the maximum SV. The η is the dynamic stress viscosity where the SV and the LSV both depend on the

- Particle size and binding forces
- Level of interactions and process of interactions
- Level of pH and size of molecules
- Capping mechanism and binding energy
- Degree of polymerization and interaction between similar molecules
- Interactions between dissimilar molecules and coalescence and coagulation
- Hydrogen bonding (HB) and electrostatic poles

Friccohesity is a natural fundamental parameter for nanoemulsions because definite amounts of cohesive forces (CFs) and entropic activities (Brownian motion) with definite amounts of frictional forces (FFs) or intermolecular forces (IMFs), noted as heteromolecular forces, are used. The Friccohesity is far beyond the concept of being the product of CF and FF. So friccohesity is leveled and matched at an extra mile mode for an incompressible and isotropic Newtonian fluid as viscous stress is related to strain rate by this simpler equation:

$$\tau = \eta \left(\frac{\partial v}{\partial y}\right)$$

Here, τ is the shear stress (drag); η is the shear viscosity of the fluid; and $\partial v/\partial y$ is the velocity derivative, which is parallel to the direction of shear, relative to displacement in the perpendicular direction. If a fluid is incompressible and viscosity is constant across the fluid, the equation is written as under:

$$\tau_{i,j} = \eta\left(\frac{\partial v_i}{\partial x_j} + \frac{\partial v_j}{\partial x_i}\right)\frac{\partial v}{\partial y}$$

Here, x_j is the jth spatial coordinate, v_i is the fluid velocity in the direction of i axis, $\tau_{i,j}$ is the jth part of stress acting on faces of the fluid element perpendicular to i axis.

- The velocity of liquid flow and the nature of structural interactions of components of liquid mixtures are interrelated.
- The stronger the interactions, the slower is the velocity and so stronger is the opposing force F, which opposes faster flow of the liquid sample with the least velocity, as under:

$$\frac{F}{A} = \eta\left(\frac{\partial v}{\partial y}\right) \quad \text{or} \quad \tau = \eta\left(\frac{\partial v}{\partial y}\right)$$

It takes longer to reach from one point to another with a low velocity, with the smallest distances toward the y axis causing $\partial v/\partial y$ of the least value.

Chapter 47

Kinematics Viscosity

Kinematics viscosity as classical mechanics deals with the motion of points or particles, bodies, and systems of bodies without taking into consideration the masses and forces of them responsible for causing motion. Kinematics depicts the geometry of the motion of a system and initial conditions of known values of position, velocity, and/or acceleration of various points that are a part of the system. Similarly, molfomatics deals with molecular forces responsible for integrating overall forces for fluid flow and developing friccohesity. Kinematics explains mechanical engineering, robotics, and biomechanics; kinematics explains the motion of systems composed of joined parts (multilink systems), like an engine, a robotic arm, or the human skeleton. One also defines total stress (tensor) that combines shear stress with conventional (thermodynamic) pressure. The equation of shear stress versus shear strain then becomes most predictive about the structural homogenization in a specific medium. Developing a nanoemulsion of suitable base pairs to repair DNA base pairs involves structural chemistry over a disruption, and both energy and activity of the molecules need optimization as follows:

- Kinetic and soothing
- Percolating and approaching

Survismeter: Fundamentals, Concepts, and Applications
Man Singh and Sunita Singh

ISBN 978-981-4774-70-3 (Hardcover), 978-0-429-02761-1 (eBook)
www.panstanford.com

- Less heating for vegetable cooking or pulse cooking
- Less vapor formation and more heat holding
- More density and more flowing and streamlining
- More mobile and use of oil for more calories

The integrational as well as the differential modes of energy, mass, and activation are summarized as:

$$\int_{x=0}^{x} x\,dx = \ln dx \quad \text{and} \quad \int \frac{1}{x} dx = d \ln x$$

If $P(n)$ depicts partitioning of the total number of particles noted as N into n then their population is calculated as under:

$$P(n) = \frac{N!}{n!(N-n)!} p^n q^{N-n}$$

The equation is defined for three different conditions where for conditions 1 and 2 could be normalized when the entire N is completely distributed and the N becomes n.

$$P(n) = \frac{N!}{N!} p^n q^{N-n} \quad \text{or} \quad P(n) = \sum_{n=0}^{n=n} p^N q$$

Such distribution patterns of a nanoemulsion could act as photosensitive materials. Thermodynamic functioning of the particles W_n is noted as under:

$$W_n = \frac{N!}{n!(N-n)!}$$

47.1 Friccohesity of SO_x, CO_x, and NO_x Gases

Gas molecules in air do not have much cohesive forces (CFs), so they could be brought together by developing their interactions with suitable solvents or a solid adsorbent. Thus, the pollutant always remains distributed in the air. The molecules of gases like SO_x, CO_x, and NO_x could be trapped by bringing them together in a most effective solvent where they are substantially dissolved. The gas molecules act as active particles, but when they are brought in contact with a solvent like water they develop a chemical reaction to form a valuable product.

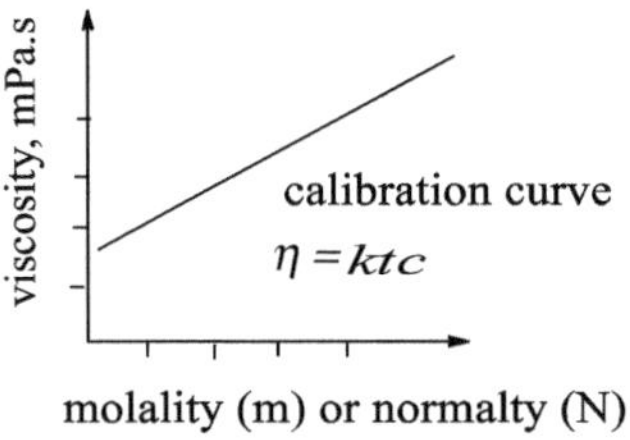

Figure 47.1 Standard calibration curve for viscosity versus composition of H_2SO_4 for determining basicity of mineral acids.

***Experiment*:** Determine the quality of air sample using the survismeter on passing the air sample into the reservoir bulb (RB) of the survismeter.

***Theory*:** The septa are used to prevent the gas sample from going out of the reservoir, and the formation of H_2SO_4 could be determined as per the reaction noted as under:

$$H_2O + SO_3 = H_2SO_4,\ \Delta H = \pm xy$$

The viscosity or the friccohesity is directly related to H_2SO_4 form on absorption of SO_3 gas, which could be extended to other gases, like CO_x and NO_x. The x stands for 1, 2, and 3. A calibration curve with known amounts of H_2SO_4 at a specific temperature is obtained using 0.005, 0.010, 0.015, and 0.020 m/L or N (normality) (Fig. 47.1).

The following equations are correlated to compositions of liquid mixtures where the mass flows and light passage in liquid mixtures could be compared as under:

$$\eta_r = \frac{\eta}{\eta_0} \quad \text{and} \quad T = \frac{I}{I_0}$$

Both η_r as relative viscosity and T as transmittance of emitted light depend on the structural interactions and could be correlated. T is the reverse of absorbance A.

$$\frac{1}{T} = A = \log_{10}\frac{1}{(I/I_0)} \quad \text{or} A = \log_{10}\left(\frac{I_0}{I}\right)$$

$\log 10 = 1$, so the equation could be modified as under:

$$A = \log_{10}\left(\frac{I_0}{I}\right)\log 10 = 1, \quad \text{and} \quad A = \left(\frac{I_0}{I}\right) \quad \text{or} \quad A = \varepsilon l\, c$$

Similarly, the penetration of light could be studied as under:

$$-\frac{dI_x}{I_x} = \frac{\sigma c A}{A} dx \quad \text{or} \quad \int_{I_0}^{I} \frac{dI_x}{I_z} = -\int_{0}^{x} \sigma c dx$$

Solving the equation, the following equation is obtained:

$$\ln(I) - \ln(I_0) = -\sigma c x \quad \text{or} \quad \ln\left(\frac{I}{I_0}\right) = -\sigma c x$$

Also the equation is resolved as under:

$$\frac{I}{I_0} = e^{-\sigma c x} \quad \text{or} \quad I = I_0 e^{-\sigma c x} \quad \text{and} \quad \eta = k t c$$

Chapter 48

Friccohesity Titrations for Hydration and Dehydration

The survismeter could be extended to other applications, like the reaction of $CuSO_4 \cdot 5H_2O$ to moderate hydration and dehydration. The reaction with copper (II) sulfate demonstrates the dehydration property of sulfuric acid. The blue crystal is changed into a white powder as water is removed. Therefore, stronger dehydrating or water as a solvent could be used. If 5 water molecules are removed from copper sulfate, the sample could be colorless. Thus the fluid dynamics of hydrated and dehydrated $CuSO_2$ could be explored with the survismeter.

$$\underset{\text{blue}}{CuSO_4 \cdot 5SO_3} \overset{H_2SO_4}{=} \underset{\text{white}}{CuSO_4} + 5H_2O, \ \Delta H = \pm xy$$

Sulfuric acid acts as a dehydrant and takes away water molecules. The resultant liquid mixtures become white or colorless. Initially, $CuSO_5 \cdot 5H_2O$ with dilute acid could be taken in the reservoir bulb (RB) and concentrated sulfuric acid could be added in lots. It shows a color change on completion of the reaction. For each addition friccohesity is measured. Such systems are noted as friccohesity titrations because the cohesive force (CF) is weakened and the water molecules get free on each addition as the friccohesity changes

Survismeter: Fundamentals, Concepts, and Applications
Man Singh and Sunita Singh

ISBN 978-981-4774-70-3 (Hardcover), 978-0-429-02761-1 (eBook)
www.panstanford.com

every time. A point comes when the friccohesity becomes a constant, which indicates the completion of water removal.

Shear stress and shear strain affect blood and skin when you use medicine and skin lotions or skin creams or facial creams. The effect could be estimated by friccohesity data. The biocompatible materials are spread on the upper surface of the skin; on spreading rheological impacts are caused. In plants, the roots could moderately attract adequate nutrient amounts from air and the liquid medium, both of which facilitate faster growth of vegetable plants. Hence an airy or spongy environment could be made by fumigating the nutrients in a vessel of selected area and dimensions. Plant seeds could be grown that cause less resistance to the roots. Structurally, the viscosity of a Newtonian liquid does not depend on stress, so the gases, water, and other liquids are treated as Newtonian in ordinary conditions but non-Newtonian fluids deviate from Newtonian liquids. Figure 48.1 shows a real state of relation of the structural states of the material, which is reflected by their shear–stress versus velocity-gradient plots. Figures 48.1 to 48.3 depict the

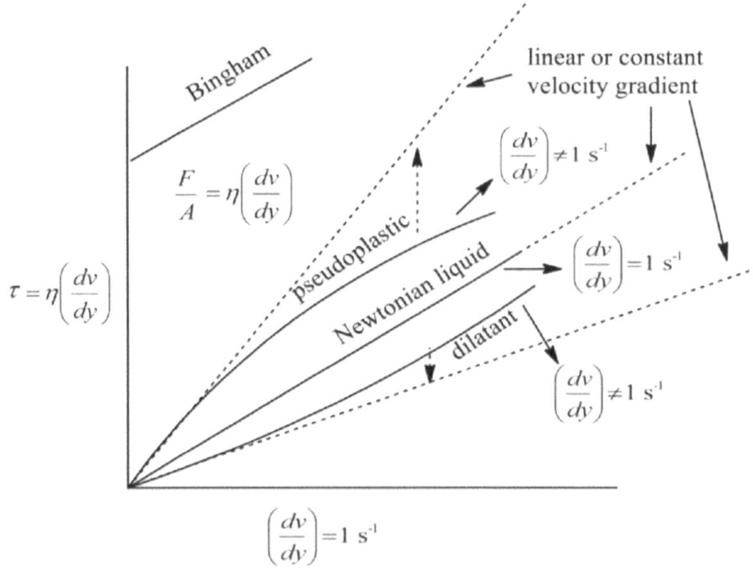

Figure 48.1 Fundamentals of fluid dynamics and states of their shear–stress rates depicting Newtonian and non-Newtonian liquids samples.

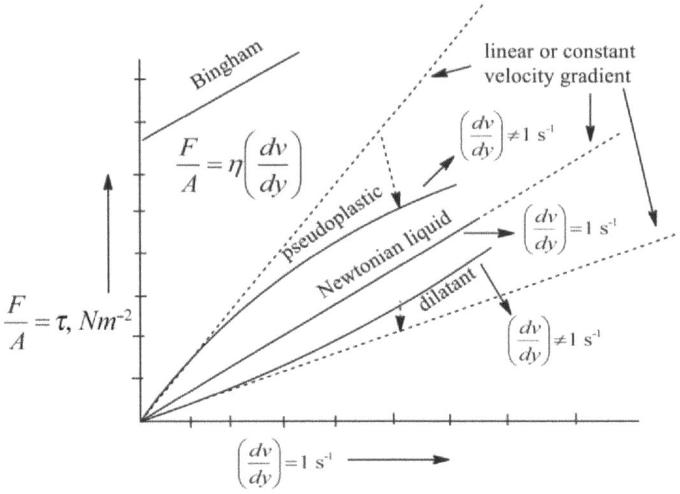

Figure 48.2 Shear stress versus shear–stress rates (dv/dy, s^{-1}) for the samples of various natures.

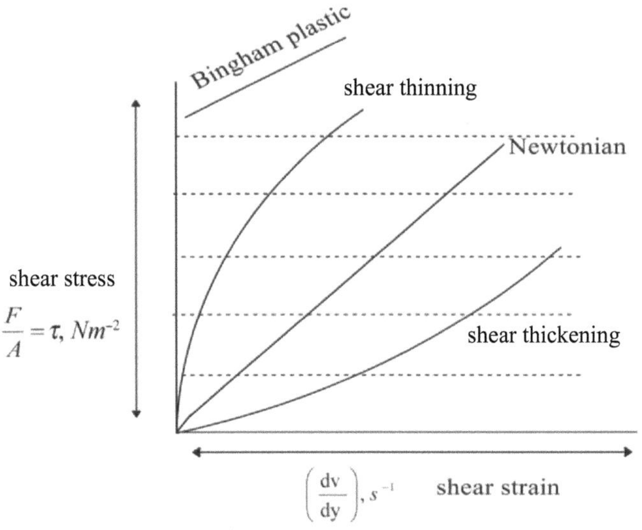

Figure 48.3 Fluid dynamics for microfluidics to determine shear thinning and thickening with decreasing and increasing viscosities with respect to the Newtonian liquid samples, respectively.

profiles of the various types of liquids. The figures depict no change in the viscosity of Newtonian liquids. Newtonian liquids do not undergo any structural change except for developing the interaction with the same rate. The dilatant increases the viscosity because the shear stress is increased, but the pseudoplastic further decreases the shear stress and the viscosity is decreased.

Also the changes in shear stress with respect to time are noted, which explain the dilatant pseudoplastic material.

The Bingham plastic exceptions show higher viscosity and friccohesity contrary to the lower surface tension and conductance. The viscosity increases and the surface tension decreases when there is shear thickening because the flow rate is reduced and a similar mass transfer takes a longer time. Hence, fundamentally, the combinatorial interfaces of the cohesion and distribution forces noted as centripetal and centrifugal forces, respectively, evolve mutual forces theory noted as friccohesity. Thus friccohesity is the most fundamental molecular science that deals with the vectoral factors of the molecular force emerging from orbital or wave functions. In a nutshell it could be inferred that cohesion or the CFs are indistinguishable forces (IDFs) whereas the forces that operate between two different molecules of different electronic configurations, like water and dimethyl sulfoxide (DMSO), are distinguishable forces (DFs). Shear thickening where stronger DFs are generated develop more and more homogenization of molecular forces, but shear thinning is just the reverse of DFs and is with stronger IDFs.

Chapter 49

Shear Thickening and Shear Thinning

49.1 Shear Thickening

The viscosity increases with the rate of shear strain because the opposing forces being applied in the unit cross area (A cm^2) are increased. When the shear strain rate becomes the least, much mass is transferred in unit time. In fact, it takes longer time in flow, noted as the viscous flow time. The opposing forces are increased because stronger molecular interactions are developed, which retard the faster flow of the liquid amount. If more and more strain is generated then the flow further becomes difficult and is noted as a thickening.

49.1.1 Shear Thickening of Eye Lotions

The thickness of eye lotions has a great impact on the eye lenses because the fluid could cause friction and hence moderate friccohesity is needed. Under a dilatant or shear thickening material viscosity and friccohesity both increase with the rate of shear strain, like a non-Newtonian fluid. This behavior usually occurs in colloids or nanoemulsions. A dilatant fluid with higher shear viscosity applies a shear stress that is a deviation from Newton's

Survismeter: Fundamentals, Concepts, and Applications
Man Singh and Sunita Singh

ISBN 978-981-4774-70-3 (Hardcover), 978-0-429-02761-1 (eBook)
www.panstanford.com

law. Such deviations are controlled by the particle size, shape, and distribution. The higher the particle size, the lower is the dilatancy, which depends on van der Waals forces and is stabilized electrostatically or sterically. Shear thickening activity occurs when a colloid transitions from a stable state to a state of flocculation. Such activities should be prevented in eye lotions as they act in a counterintuitive way on striking against a surface. It is completely soaked with water behaves as a dilatant material with higher shear stress. Rheopecty is a similar property in which the viscosity increases with cumulative stress or agitation over time. The opposite of a dilatant material is a pseudoplastic. The shear thickening leads to higher friccohesity, enabling several applications.

49.2 Fundamentals of Dealing with a Liquid in Daily Activities

Right from drinking water to food, cosmetics, clothes washing, and industrial formulations, we deal with a variety of academic as well as industrial liquids with many physicochemical activities with several normalized states. Such physicochemical properties produce definite data, so such properties become the sensor to define the quality of the desired formulations. Thereby, the experiments and data both become important to understand their friccohesity for better formulations and products obtained in a reliable manner.

49.2.1 Academic applications of liquids

Academically, we use liquids for conducting reactions as an effective collision-inducing medium at variable temperatures and pressure, pH, and chemical additives. Cleaning glassware or clothes or other solid surfaces by dissolving or solubilizing various chemical compounds, like salts, organic compounds, surfactants, dendrimers, and other processes, is being routinely done in labs.

49.2.2 Teeth-cleaning biokinetic liquids

These are formulated from natural resources that are nanoemulsions with strong kinetic energy and friccohesity. These could enter the interteeth gaps.

49.2.3 Industrial applications of liquids

Liquids are used for intermixing of chemicals, dye dissolving for thin film formation, clothes washing, dry cleaning, surface cleaning, agrochemical and pesticides, ink making, and making of food and nutrition items. Therefore, we come across many situations in which liquid materials are used.

49.2.4 Interfacing between FTIR and friccohesity

Molecular populations in the dispersed medium matter a lot because the larger the number of molecules, the stronger are the binding forces where the stretching frequency is affected as the functional groups or the atoms that were comparatively strongly bound on stronger interactions are less vibrant. Thus the friccohesity of the material is also influenced by the cohesive forces and frictional forces. Thus functional interfaces could be developed.

49.3 Shear Thinning

The viscosity decreases with the rate of shear strain because the opposing forces are weakened, which allow for a quicker mass transfer as weak intermolecular forces (IMFs) are developed. Thereby in the absence of weaker forces, within the minimum time the maximum mass is transferred. It also has stronger surface tension because due to stronger cohesive forces (CFs) the mixture cannot develop stronger interactions leading to the generation of stronger IMFs. This proves that the solute and solvent interactions are weakened and at a microlevel the solvent nucleates to solvent binding while the solute nucleates to solute site. This type of interacting mechanism of solute and solvent is noted as reverse interacting activities with weaker solute and solvent interactions. Thus the process is noted as shear thinning.

49.4 Thixotropic Liquids

Such liquids become less viscous over time when stirring, oscillated, or stressed. Since the mixtures have weaker IMFs and chemical activities, a shaking activity provides mechanical energy that further weakens the exiting IMF, so their viscosity decreases.

Chapter 50

Rheopectic Liquids

Rheopectic liquids become more viscous over time when shaken, agitated, or stressed.

50.1 Bingham Plastics

These behave as solids at low stress but flow as viscous fluids at high stress. For a Newtonian fluid, the viscosity depends on its composition and temperature but for gases and other compressible fluids, it depends on temperature and varies very slowly with pressure. A magnetorheological fluid, for example, becomes thicker when subjected to a magnetic field, possibly to the point of behaving like a solid. The thickening activities weaken the cohesive forces (CFs), which increases the number of pdn. The pendant drop involves both adhesive forces and CFs. Near a solid surface area the system has adhesive forces, whereas at a slight distance from the solid surface the liquid–liquid molecules develop CFs at the cost of Gibbs energy, noted as under:

$$\Delta G = RT \ln \left(\frac{dv}{dy}\right)_{\mathrm{T,P}} \quad \text{and} \quad \Delta G = RT \ln K_{\mathrm{eq}}$$

Survismeter: Fundamentals, Concepts, and Applications
Man Singh and Sunita Singh

ISBN 978-981-4774-70-3 (Hardcover), 978-0-429-02761-1 (eBook)
www.panstanford.com

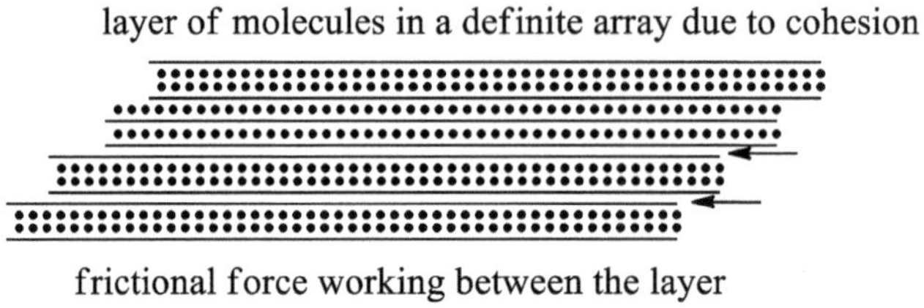

Figure 50.1 Ideal laminar flow of liquid samples with a definite sheet dimension and friccohesity generating well-defined velocity gradients with perfectly constant viscous coefficient.

Thus dv/dy could be determined selecting a series of solvents of known viscosities and the systems remain in equilibrium, which is depicted as under:

$$\ln \eta_{eq} = \ln\left(\frac{dv}{dy}\right) \quad \text{or} \quad \ln K_{eq} = \ln\left(\frac{dv}{dy}\right)$$

The equilibrium is between laminar layers causing frictional forces (FFs) despite cohesion between the molecules of each layer, as is depicted in Fig. 50.1.

The preferential interacting sites lead to a tentropy because the geometrical shape and steric spatial arrays with respect to the order of flow set the patterns of flow. The tentropy could be determined using the following equation:

$$\text{Tentropy} = \Delta E^* \cdot n_b \cdot \sigma_b \cdot n(-CH_2-) \cdot n(CH_3-) \cdot n_{fg} \cdot \mu \cdot \sigma_f.$$

Here ΔE^* is the activation energy, n_b is the number of nonbonding electron pairs, σ_b is the number of sigma bonds, $n(-CH_2-)$ is the number of methylene groups, $n(CH_3-)$ is the number of methyl groups, n_{fg} is the number of functional groups, and μ is the dipole moment of the solution in a specific solvent system. Several times many turbulences work that cause many flow rates within the layers where the solvent binds with the main particles or molecules distributed inside the solvent. So solute and solvent engineering causes tentropy versus friccohesity expressions and the hydrodynamic volume is generated, which is determined using the following equation:

$$\text{hydrodynamic volume} \times \text{dipole moment} = \tau\left(\frac{\eta}{\eta_0} = \eta_r\right)$$

Hydrodynamic volume and molecular size are depicted as under:

$$\text{hydrodynamic volume} = [\eta]M$$

Hydrodynamic volume is the total volume of the particles at the core and the solvent volume making a layer. The partial CFs and partial locally entropic forces work and the CFs as well as entropic forces are noted as friccohesity. Thereby, the general equation is corrected as under:

$$\text{hydrodynamic volume} = [\eta]M[\sigma_r]$$

Also the size of the molecules is determined because the M varies so the size varies.

$$[\eta] = kM^a$$

Here, $M = \text{R–O–SO}_3\text{Ns}$ and R changes from $C = 8$ to 20 so the M variation is determined. If the a is zero then the system is ideal as $M^0 = 1$ and $\log 1 = 0$.

50.2 Foamy Mixture

Both UV and viscosity directly depend on molecular interactions, which are derived by varying the compositions of the liquid mixture. Therefore, this theme could be applied for developing the lightest liquid filled with foam, air, and high-class surfactants so that the resultant mixtures have the lowest binding forces. Thereby, the formulations could replace the use of temperature-based supercritical liquids. New possible formulations could be developed by mixing butane and pentane gases.

- Butane gas is a highly flammable, colorless, easily liquefied gas.
- Pentane is an alkane with 5 carbon atoms, *n*-pentane isomer, isopentane (methyl butane), and neopentane (dimethyl propane). Cyclopentane is not an isomer of pentane because it has only 10 hydrogen atoms whereas pentane has 12 C.
- Pentanes are components of fuels and employed as specialty solvents in laboratory.
- Pentanes are some of the primary blowing agents used in the production of polystyrene foam and other foams.

- Pentane's low boiling point, low cost, and relative safety allow its use as a working medium in geothermal power stations in some blended refrigerants.
- Pentanes are an active ingredient in some pesticides.

In light of above the following formulations are used:

Butane + pentane + surfactant + ethanol + water + CO_2+ O_2+ dendrimer + MFP resin + methane + propane + TMSOI + MTOIC + oil + waste mobile oil + kerosene oil

The said formulation could be a most usable supercritical fluid (SCF) in industries. Such materials could be studied with UV absorbance. UV absorbance and viscosity both are the function of compositions, so the data of both properties could directly illustrate the chemical activities occurring in a process. Along with size-based interactions the surface area of the molecules, which is determined with the survismeter, could indicate the quality of foamy mixtures. Both UV and viscosity are competent properties that elucidate the mechanism of interactions like drug binding or release activities.

50.3 Foamy Medium for Airy Tree Growth

A foamy medium could be most useful for supporting airy tree growth by providing a porous structural framework or a soft nylon base or metallic scrubbers for supporting the vertical growth of

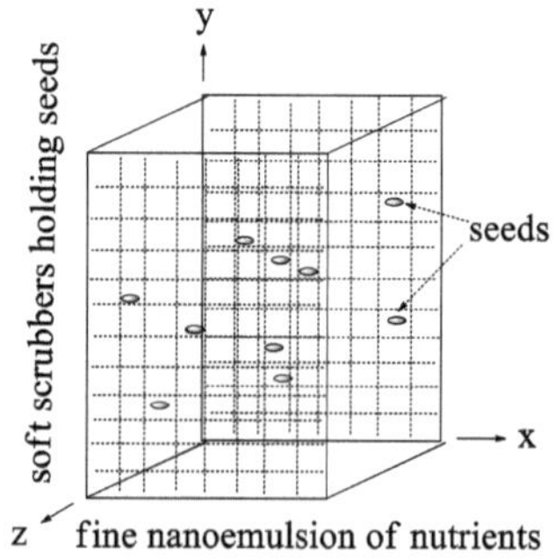

Figure 50.2 An excellent model of nanoemulsions supporting growth of several vegetable plants because such nanoemulsions are aerated and mixed with plant nutrients.

the plant while allowing absorption of nutrients from the medium. Certain vegetables could be grown whose plant height is not much. This approach could be the best one for making the best use of waste organic materials and wastewater. Such foamy or airy nanoemulsion must be thermodynamically and kinetically stable so the medium does not coagulate or coalesce with time. Such kinetics must be avoided. The hydrophobicity component is needed for such foamy emulsions because air could be trapped by them similar to the air balloon. Froth stability is the most desired element in such cases. Therefore, detergents, soaps, or soft polymers of higher hydrophobicity could be used in specific ratios. Similarly, air-trapping capillaries could be generated by choosing a surfactant with a longer alkyl chain. So the contribution of the alkyl chain becomes useful and a need arises to determine the number of $-CH_2-$ groups in the molecules.

Chapter 51

Contribution of the Number of $-CH_2-$ Groups

Determine the contribution of the number of $-CH_2-$ groups in cationic, anionic, and nonionic surfactants. Also their variable structures with definite hydrophilic and hydrophobic components are developed in the prescribed medium. The structures given in Fig. 51.1 could also assist in the development of a porous or airy medium.

In such processes substantial chemical forces work and hence transformation from the cohesive forces (CFs) of the solvent to the intermolecular forces (IMFs) of liquid solutions is noted as under:

$$-\frac{d(a)}{dt} \propto (a - x) \quad \text{Or} \tag{51.1}$$

$$-\frac{d(a)}{dt} = k(a - x) \tag{51.2}$$

Let a be the CF of the solvent and x stand for IMF. Similarly, the CF depicts indistinguishable forces (IDFs) and distinguishable forces (DFs).

$$-\frac{d(\text{CF})}{dt} \propto (\text{CF} - \text{IMF}) \quad \text{or} \tag{51.3}$$

$$-\frac{d(\text{CF})}{dt} = k(\text{CF} - \text{IMF}) \tag{51.4}$$

Survismeter: Fundamentals, Concepts, and Applications
Man Singh and Sunita Singh

ISBN 978-981-4774-70-3 (Hardcover), 978-0-429-02761-1 (eBook)
www.panstanford.com

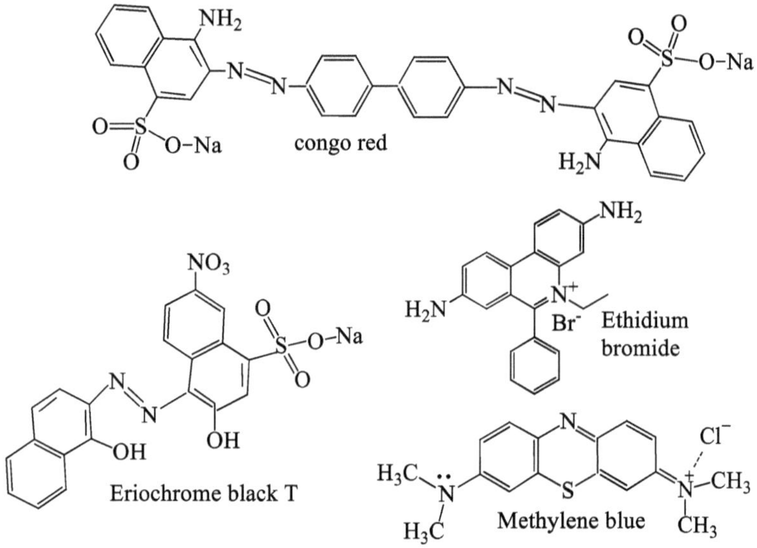

Figure 51.1 Molecular structures with different interacting sites inside their structural frameworks, which initiate philic–phobic activities for industrial formulations due to sharper tentropy of these molecules.

On rearranging similar terms together Eqs. 51.1 and 51.2 are noted, which also define the limits from 0 to t time.

$$-\frac{d(a)}{(a-x)} = kdt \quad \text{or} \tag{51.5}$$

$$\int_0^t \frac{d(a)}{(a-x)} = -k\int_0^t dt \tag{51.6}$$

The integration is solved as under:

$$\ln(a_t - a_0) = -kt + I \quad \text{and} \tag{51.7}$$

$$\ln a_t - \ln a_0 = -kt + I \tag{51.8}$$

The equation is solved as under:

$$\ln\frac{a_t}{a_0} = -kt + I \quad \text{or} \tag{51.9}$$

$$\frac{a_t}{a_0} = e^{-kt} \quad \text{or} \quad a_t = a_0 e^{-kt} \tag{51.10}$$

The alternative derivation is as under:

$$-\frac{d(a)}{dt} \propto (a-x) \quad \text{or} \tag{51.11}$$

$$-\frac{d(a)}{dt} = k(a-x) \tag{51.12}$$

Also the equation is noted as under:

$$-\frac{d(a)}{(a-x)} = kdt \quad \text{and} \tag{51.13}$$

$$\int_0^t \frac{d(a)}{(a-x)} = -k\int_0^t dt \tag{51.14}$$

$$\ln(a-x) = -kt + I \tag{51.15}$$

At $t = 0$ time, $x = 0$ and $a = a_0$. Putting these values in Eq. 51.5, we get Eq. 51.6.

$$\ln(a_0 - 0) = I \quad \text{or} \quad \ln(a_0) = I \tag{51.16}$$

On putting the value of I from Eq. 51.6 into Eq. 51.5, a new equation, Eq. 51.7, is obtained.

$$\ln(a-0) = -kt + \ln(a_0) \tag{51.17}$$

On rearranging Eq. 51.7, we get Eq. 51.8.

$$\ln(a-x) - \ln(a_0) = -kt \tag{51.18}$$

Changing the minus sign in Eq. 51.7,

$$-\ln(a-x) + \ln(a_0) = kt \quad \text{or} \quad \ln\frac{a_0}{(a-x)} = kt \tag{51.19}$$

$$\text{or} \qquad k = \frac{2.303}{t}\log\frac{a_0}{(a-x)} \tag{51.20}$$

The exponential form of Eq. 51.7 is developed as under:

$$\ln\frac{(a-x)}{a_0} = -kt \quad \text{or} \quad \frac{(a-x)}{a_0} = e^{-kt} \tag{51.21}$$

Let us apply the kinetics of friccohesity.

Let at $t = m$ mol/L, $x = \text{IMF}$, and $a = \text{CF}$

$$k = \frac{2.303}{m}\log\frac{\text{CF}_0}{(\text{CF}_0 - \text{IMF})} \tag{51.22}$$

On putting the values of $CF_0 = \gamma$ (surface tension of the solvent) and $IMF = \Delta\eta$, Eq. 51.11 becomes Eq. 51.12.

$$k_{\text{friccohesity}} = \frac{2.303}{m} \log \frac{\gamma_0}{(\gamma_0 - \Delta\eta)} \tag{51.23}$$

Equation 51.12 estimates friccohesity, which could act as a constant $k_{\text{friccohesity}} \cdot \gamma_0$ is the surface tension of the medium acting as a solvent, and $\Delta\eta$ depicts a change in viscosity with m mol/kg solute.

$$k_{\text{friccohesity}} = \frac{2.303}{m} \log \frac{\gamma_0}{(\gamma_0 - \Delta\eta)} \tag{51.24}$$

Conditions

If $\Delta\eta = 0$, then Eq. 51.12 becomes

$$k_{\text{friccohesity}} = \frac{2.303}{m} \log \frac{\gamma_0}{(\gamma_0)} \tag{51.25}$$

$$\text{or} \quad k_{\text{friccohesity}} = \frac{2.303}{m} \log 1 \tag{51.26}$$

$$k_{\text{friccohesity}} = 0 \tag{51.27}$$

The value $\Delta\eta > 0$ induces $k_{\text{friccohesity}} > 0$, which indicates strengthening of mutual mixing.

Chapter 52

Friccohesity of Micelle Formation

Since all the fragmented surfactant molecules come together because like attracts like, cohesive forces (CFs) are created as follows:

- Formulation of friccohesity of the cage model of reaction
- Hair adhesives and liquid soaps
- Dyeing fabrics and hydrodynamics

In soil due to CFs the H_2O gets accumulated in large numbers, which is available for the roots of plants. Two bonding adhesives become operational on spreading and developing adhesive forces: solvent based and water based. Water develops adhesive forces with the soil particles and the roots enter the soil in search of the water, as is noted in the Fig. 52.1.

When a liquid maintains contact with a solid surface via intermolecular interactions to bring the two surfaces together, it is known as wetting. The level of wetting is determined by the force balance between adhesive forces and CFs, which has become significant in nanotechnology and nanosciences due to the use of nanomaterials. Wetting is useful in bonding or adherence of two materials; the wetting and surface forces cause capillary effects.

Survismeter: Fundamentals, Concepts, and Applications
Man Singh and Sunita Singh

ISBN 978-981-4774-70-3 (Hardcover), 978-0-429-02761-1 (eBook)
www.panstanford.com

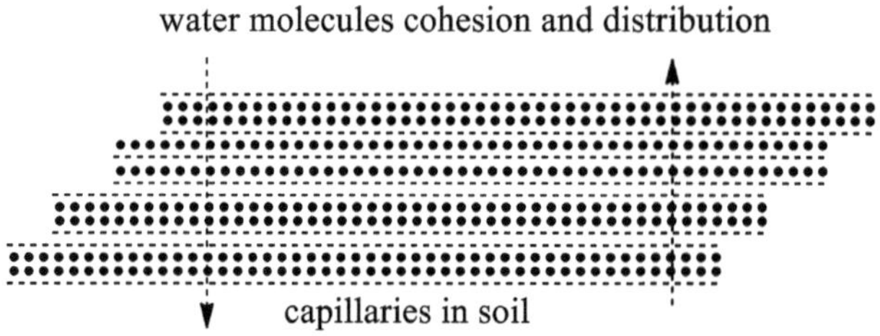

Figure 52.1 Microfluidics of a perfectly mixed or homogeneous liquid sample with higher friccohesity and lower surface energy.

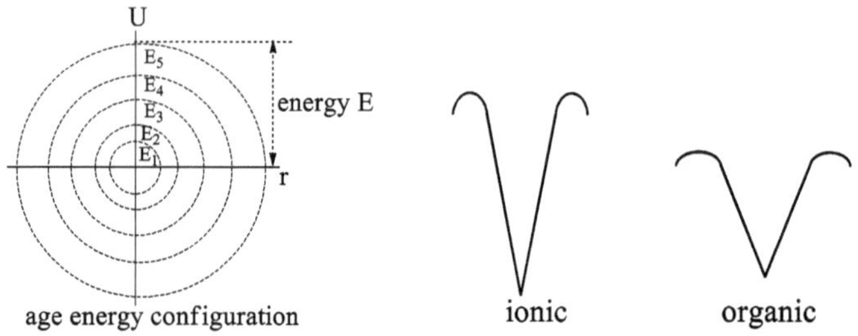

Figure 52.2 Configurations of energetics of liquid sample formulations of ionic species and organic molecules with sharper and loosened impacts of potential energy states.

The cage energy formation on interaction with ionic and organic molecules is noted in Fig. 52.2.

U is the potential energy, r is the distance, and E is the energy. The cage energy or the energy intensity in the case of ionic is sharper and for organic is less sharp due to shared sigma bonds, which occupy larger surface energy, which is explained by the curvature, slope, and depth of the cage, as is noted in Eq. 52.1.

$$\text{curvature} = \frac{\partial^2 U}{\partial r^2},\ \text{slope} = -\frac{\partial U}{\partial r},\ \text{depth} = \mathrm{U_{max}} - U_{\mathrm{mini}} \qquad (52.1)$$

As compared to pentaerythritol tetranitrate (PETN) and cyclotrimethylenetrinitramine (cyclonite) the triglyceride with variable fatty acids has higher stability and less density due to larger numbers of sigma bonds, which occupy a larger volume. PETN and cyclonite with several charged points develop coulomb attraction. The physicochemical interactions of PETN could be investigated

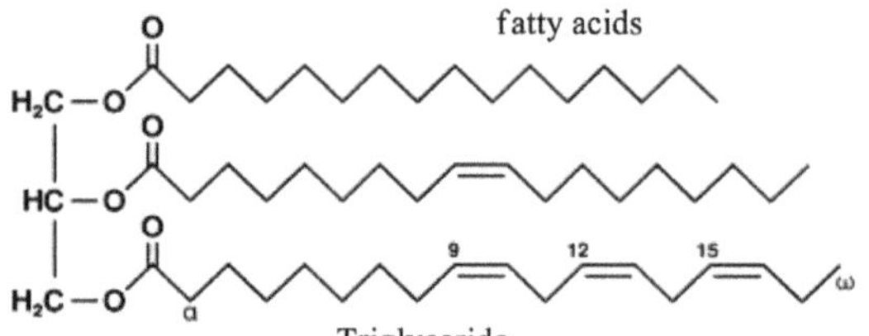

Figure 52.3 Structural representation of triglycerides (oils) having different fatty acids with various chemical bonds responsible for degree of saturation of oils, hydrolysis and their nanoemulsion formations.

in several solvents to check its interacting abilities. The study of friccohesity of such explosives in polar, apolar, protic, and aprotic polar solvents could an interesting hypothesis for understanding the mechanism of the negatively and positively charged multipoles. Such molecules could also assist in the development of nanoemulsions of oil and water and water and oil, both with thermodynamic and kinetic stability. The formulation of oil/water nanoemulsions with PETN and cyclonite is estimated with the survismeter. PETN could develop a nanoemulsion of heavy oils and water as there is difficulty regarding how to discard the heavy oils. The interactions of PETN with cationic, anionic, and nonionic surfactants could be investigated for micelle forming activities. PETN could also activate water molecules to percolate in a better way with the hydrocarbon chains of the surfactants and oils. Also the air-trapping ability of the surfactants could be studied with PETN. The nanoemulsion of the PETN could act as a controlled fuel for generating the energy because PETN has a lot of oxygen. PETN and starch paste could also be used as fuels, which should be carefully used and small amounts of PETN should be used.

Structurally the viscosity of oil as compared to 0.8903 mPa·s of water is 57 mPa·s but the surface tension and density both are lower than that of water. In oil the hybridization of electrons within the shared orbital covers a larger surface area as compared to ionic charges that attract toward a center. This type of approach of electronic combinations induces stronger ionic charges with a comparatively higher density. The point charges develop higher density and surface tension, contrary to a state where the electrons

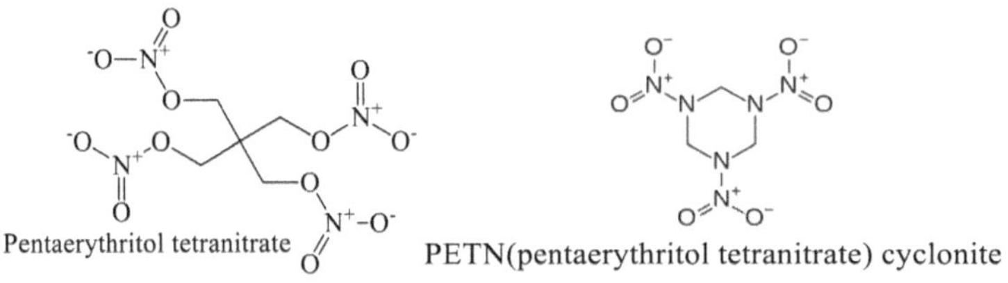

Figure 52.4 Structural constituents of highly active molecules to be explored to catalyze the hydrolysis of triglycerides (oils) and their nanoemulsion formulation.

are spread in specific areas of molecules. In light of the saturated and ionic charges a few queries could be resolved by conducting the following experiment using the survismeter:

(1) Determine saturated and unsaturated oils with friccohesity.
(2) Determine the percentage of cancer causing oils, since saturated oils cause cancer.
(3) Determine the dissolution of cholesterol by a given solvent system.
(4) Determine the dissolution of pesticides for better and wise use of pesticides.
(5) Determine the presence of jaundice on the basis of the urine sample of a person.
(6) Determine the purity of table salts and glucose.
(7) Determine the purity of a water sample with the survismeter.
(8) Determine the intercalation of DNA by silibinin as a drug molecule.
(9) Determine the binding of a drug complex.
(10) Determine the drug release rate from a drug-vehicle complex.
(11) Determine the efficiency of soaps and detergents.
(12) Determine the paint removing ability of a solvent or a surfactant.
(13) Determine the critical micelle concentration (CMC) of a surfactant in a specific set of solvents.
(14) Determine the adsorption of a drug or an acid by activated charcoal.
(15) Determine the alkali content of a soap sample.
(16) Determine the ionic field of table salt.

(17) Determine the current passing ability of salts.
(18) Determine free radical scavenging by a dendrimer sample.
(19) Determine the mixing ability of two samples.
(20) Determine the water adsorbing ability of an adsorbent.
(21) Determine sol gel formation by a visionmeter.
(22) Determine the conducting ability of a thin film.
(23) Determine the water removing ability of hydrophobes.
(24) Partition energy in nanoemulsions.

Chapter 53

Variance in Data

Variance (denoted by σ^2, s^2, or var.) is a squared deviation of a random variable from its mean. It measures the spreading of random numbers from their mean. It tests the concepts of physics, biology, and chemistry. Variance is a square of standard deviation, the second central moment of a distribution, and covariance of a random variable with itself, depicted as under:

$$\sigma = \sqrt{\frac{(x-\mu)^2}{n}} \quad \text{or} \quad \sigma^2 = \frac{(x-\mu)^2}{n}$$

Therefore, standard deviation is the scaling of a variable.

53.1 Probability Density Function

Probability density function (PDF) or densities of a continuous random variable or function define a sample or point in the sample space as a relative likelihood. The absolute value of a continuous random variable is 0. For infinite sets of possible values, the PDF at two different samples infers a random variable as under:

$$f(x, \mu, \sigma^2) = \frac{1}{\sigma\sqrt{2\pi}} e^{-\frac{1}{2}\left(\frac{x-\mu}{\sigma}\right)^2}$$

Survismeter: Fundamentals, Concepts, and Applications
Man Singh and Sunita Singh

ISBN 978-981-4774-70-3 (Hardcover), 978-0-429-02761-1 (eBook)
www.panstanford.com

For a standard normal distribution $\mu = 0$ and $\sigma = 1$ in the equation. The x data points or sample points develop probability density $(p = \phi)$.

$$f(x) = \frac{1}{\sqrt{2\pi}} e^{-\frac{1}{2}x^2} \quad \text{or} \quad p(x) = \frac{e^{-\frac{1}{2}x^2}}{\sqrt{2\pi}}$$

$1/\sqrt{2\pi}$ ensures a total area under curve $\phi(x) = 1$ and the value $\frac{1}{2}$ controls the distribution to have a unit variance $(\sigma^2 = 1)$ or a standard deviation. This function is symmetric around $x = 0$, and $1/\sqrt{2\pi}$ attains its maximum value with inflection points at $+1$ and -1. When $\sigma^2 = \frac{1}{2}$ then $p(x)$ is noted as under:

$$f(x) = \frac{1}{\sqrt{\pi}} e^{-x^2}$$

53.2 General Normal Distribution

Normal distribution is a standard normal distribution stretching its domain by σ (deviation) and translated by μ value.

$$f(x, \mu, \sigma^2) = \frac{1}{\sigma} p(\frac{x - \mu}{\sigma})$$

The PDF is scaled by $1/\sigma$ for getting integral $= 1$. If the standard normal deviation is Z then $X = Z\sigma + \mu$ infers a normal distribution with the expected value of μ and standard deviation σ. If X is the general normal deviation then $Z = (X - \mu)/\sigma$ infers a standard normal distribution. Both the local equilibrium in the sample volume in the buffer bulb (BB) and its physicochemical properties are related to the state variables in the same manner as in equilibrium. The power series of e^x is determined as under:

$$e^x = \sum_{n=0}^{\infty} \frac{x^n}{n!} = 1 + x + \frac{x^2}{n!} + \frac{x^3}{3!} + \frac{x^4}{n!}$$

If n is from zero to 4 and $x = 3$,

$$e^x = \sum_{n=0}^{4} \frac{x^4}{4!} = 1 + x + \frac{x^2}{2!} + \frac{x^3}{3!} + \frac{x^4}{4!}$$

Chapter 54

Internal Relaxation Process

The internal relaxation process in a fluid is faster than the rate of change due to the state variables. So a local equilibrium concept is valid for the transport and rate processes of fluid systems. The relaxation time for heat conduction by gases at normal temperature and pressure (NTP, which is 25°C and 1 atm pressure) is 10^{-12} s, contrary to 10^{-11} to 10^{-13} s for fluids. The relaxation time is much higher for a structured liquid. An extension of equilibrium thermodynamics to nonequilibrium systems with the local equilibrium is in terms of entropy $s[\mathrm{T}(x), \mathrm{ni}(x)]$ and energy $u[\mathrm{T}(x), \mathrm{ni}(x)]$ densities. The $s[\mathrm{T}(x), \mathrm{ni}(x)]$ and energy $u[\mathrm{T}(x), \mathrm{ni}(x)]$ are functions of temperature and mole number densities at location x at a local temperature T (x). Total entropy and energy are obtained from integrals over a volume of the system.

$$dE = h\nu \tag{54.1}$$

$$\text{and} \quad \nu = \frac{1}{2\pi}\sqrt{\frac{k}{\mu}} \tag{54.2}$$

By putting the value of ν from Eq. 54.2 in Eq. 54.1 we can further define dE as under:

$$dE = h\frac{1}{2\pi}\sqrt{\frac{k}{\mu}} \tag{54.3}$$

Survismeter: Fundamentals, Concepts, and Applications
Man Singh and Sunita Singh

ISBN 978-981-4774-70-3 (Hardcover), 978-0-429-02761-1 (eBook)
www.panstanford.com

dE is further connected to the traditional thermodynamic equation as follows:

$$dE = TdS - PdV + \sum_{i=1}^{n} \mu_i dn_i \tag{54.4}$$

Hence, putting the value of dE from Eq. 54.3 in Eq. 54.4, we obtain Eq. 54.5.

$$\frac{h}{2\pi}\sqrt{\frac{k}{\mu_r}} = TdS - PdV + \sum_{i=1}^{n} \mu_i dn_i \tag{54.5}$$

Equation 54.5 depicts a state of molecules that involves kinetic, dynamic, and translational motions with a definite entropy, pressure, volume change, and chemical potential. If volume change is negligible then $dV = 0$ and Eq. 54.6 is obtained.

$$\frac{h}{2\pi}\sqrt{\frac{k}{\mu_r}} = TdS + \sum_{i=1}^{n} \mu_i dn_i \tag{54.6}$$

On rearranging Eq. 54.6, dS is obtained as under:

$$\frac{h}{2\pi}\left(\frac{1}{T}\right)\sqrt{\frac{k}{\mu_r}} - \left(\frac{1}{T}\right)\sum_{i=1}^{n} \mu_i dn_i = dS \tag{54.7}$$

However, the traditional thermodynamic equation is noted as under:

$$\Delta G = \Delta H - T\Delta S \tag{54.8}$$

$$T\Delta S = TdS = dE + PdV - \sum \mu_i dn_i \tag{54.9}$$

Equations 54.8 and 54.9 are also used to develop Eq. 54.10.

$$\Delta G = \Delta H - [dE + PdV - \sum_{i=1}^{n} \mu_i dn_i] \tag{54.10}$$

Putting the value of dE from Eq. 54.3 in Eq. 54.10, the following equation is obtained:

$$\Delta G = \Delta H - dE - PdV + \sum_{i=1}^{n} \mu_I dn_i \tag{54.11}$$

$$-\ln k_{eq} = \Delta H\left(\frac{1}{nRT}\right) - \left(\frac{1}{nRT}\right)\frac{h}{2\pi}\sqrt{\frac{k}{\mu}} - \left(\frac{1}{nRT}\right)PdV + \left(\frac{1}{nRT}\right)\sum_{i=1}^{n} \mu_I dn_i \tag{54.12}$$

$$\sigma = \sqrt{\frac{(x-\mu)^2}{n}} \quad \text{or} \quad \sigma^2 = \frac{(x-\mu)^2}{n} \tag{54.13}$$

Equation 54.13 depicts variance or standard deviation. Further, the uncertainty is noted as Eq. 54.14.

$$\text{uncertainty} = \left(\frac{\sigma}{\sqrt{n}}\right) \tag{54.14}$$

The scattering in data is calculated using Eq. 54.15.

$$\text{scattering} = (x - \mu)^2 \tag{54.15}$$

The standard error of mean (σ_m) is calculated from the standard deviation noted as Eq. 54.16.

$$\sigma_m = \left(\frac{\sigma}{\sqrt{n}}\right) \tag{54.16}$$

The mean value μ on a scale of 95% or 1.96 distributions is used to calculate the lower and upper limits, as under:

$$\text{lower limit } = \mu - (1.96)(\sigma_m) \text{ and upper limit } = \mu + (1.96)(\sigma_m) \tag{54.17}$$

Contrary to the mass transfer in the case of transport properties chemical processes are conducted on the passage of electrons from one compartment to another through a charge conducting or holding object or medium. The cell has a Zn solid electrode dipped in $ZnSO_4$ and a Cu solid electrode dipped in $CuSO_4$ solutions. Due to a difference in metallic pressures of the Zn in the solid state and in the liquid state a potential difference exists that becomes a driving force for conducting the charge via the medium. However, there are certain media that do not support the passage of electrons, called nonconducting media. The salts in water become a conducting medium, but the ions, which develop a strong ionic charge and then stronger coulombic interactions than the conducting ability, become weak with low conductance. Thus the cohesive forces (CFs) that are developed between the oppositely charged water solvation spheres influence the conducting property. But under such strong CFs the friccohesity, viscosity, and surface tension increase and the charge conducting ability or the conductance decreases. For such a mechanism the friccohesity becomes the most effective data. The cell has a Zn solid electrode dipped in $ZnSO_4$ and a Pt solid electrode dipped in HNO_3 solution (Fig. 54.1). Therefore, the electric cell and survismeter sciences are interrelated in terms of friccohesity, which is more an electronic parameter rather than a physicochemical

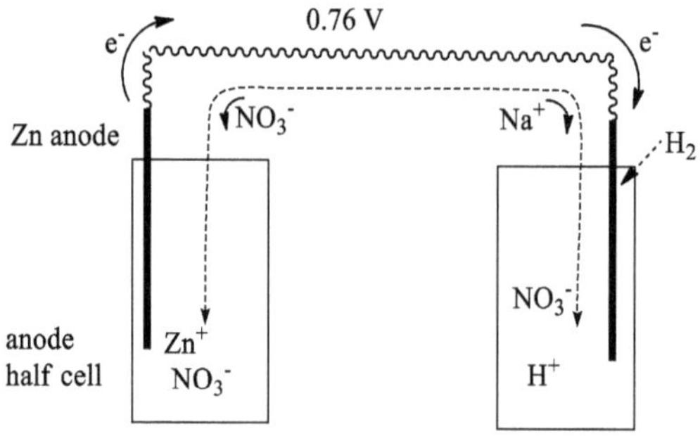

Figure 54.1 Ionic flows in continuous liquid phase toward counter electrodes that are affected by shear stress and charge spontaneity. Higher shear stress or higher friccohesity weakens their flow rates.

parameter. Hence, the friccohesity could be correlated to the electronic models of the molecules, namely the highest occupied molecular orbital (HOMO) and the lowest unoccupied molecular orbital (LUMO). For example, the stronger the electrostatic forces, the stronger are the CF and for a stronger electrostatic charge center the shared electron pair is shifted toward a more electronic atom. In such a situation the orbital becomes less occupied and so is noted as LUMO but the surface tension is higher and the friccohesity is lower. This is contrary to saturated oils, which develop nanoemulsions and produce least CF and higher friccohesity. Such nanoemulsions are noted as HOMO because stoichiometry of their ingredients is highly balanced and no residual forces are left unbalanced. Figure 54.1 depicts a potentiometric cell where the charge or the electron passage is made through the solid electrodes via the medium.

Zn metal is dissolved in $Zn(NO_3)_2$ (aq) at the anode in the voltaic equation, listed as under:

$$Zn(s) = Zn^{2+}(aq) + 2e \quad \text{and} \quad 2H^+(aq) + 2e = H_2(g)$$

H_2 is passed into HNO_3 (aq) dipped in the Pt electrode. Such devices draw current from metals when they are used as electrodes in the most relevant solution noted as electrolyte. 3-aminopropyl) triethoxysilane gets solidified in environmental studies.

$$\Delta E^* \propto \sigma \quad \text{or} \quad \Delta E^* = k\sigma \quad \text{(fundamental eqn. 54.1)}$$

Structural and physicochemical signatures of molecules lay the foundation for physicochemical properties of molecular behavior in homogenous or heterogamous systems with separate CFs. It happens due to the following points:

- Stronger hybridization and solvent mismatch
- Nonelectrostatics and columbic poles
- Medium permittivity and polarity of the mixtures
- String of units to work
- Friccohesity a mode to reduce the body weight

Chapter 55

Survismeter Science of Structure and Number of Molecules

Most closely the structure and number of molecules in a liquid sample respond sensitively to predict the chemical activities on mixing the interacting and reacting species at a prescribed temperature and pressure. Both the structure and number of molecules induce internal pressure at a prescribed temperature and volume. Such interacting mechanisms of drug molecules tend to cure diseases. Molecular collisions in Lennard-Jones potential (LJP) distribution and their potential energy both speak out their interatomic distance. The dielectric forces (DEFs) as a zonal potential that integrates the molecular potential form a solution that leads to speak out the molecular pattern. Thus, the cell functioning occurs due to standard friccohesity with an adequate cohesion-adhesion-IMF (intermolecular force) chain for needful entropy within the living and nonliving boundaries. Particles having a water-binding capacity of the soil constitute structured fluid dynamics driven by the friccohesity, which senses the pressure changes inside the solution. Since the measurement of surface tension and viscosity separately by devices is not accurate, approaches that use separate measurements contaminate. Hence friccohesity

Survismeter: Fundamentals, Concepts, and Applications
Man Singh and Sunita Singh

ISBN 978-981-4774-70-3 (Hardcover), 978-0-429-02761-1 (eBook)
www.panstanford.com

measurements integrate whether salts or surfactants are present in mixtures. Similarly, dimerization, coagulation, and coalescence are involved and these could be studied with the survismeter, which could be applied for samples containing paints and plastic. So a friccohesity sensor could be effective, which may also be verified by dipping the infrared sensor in the reservoir bulb (RB). Such approaches could sense the quantity of crystal or degeneration by measuring the emitting or loading light intensity.

55.1 Designing of New Nanoemulsions

Formulate nanoemulsions out of hexanol, trimesoyl tridihexyl malonate ester (TTDHM) ethanol + water, propanol + water, and heptanol + water as compared to pure petrol or diesel.

The approach could be extended for the identification of saturated oils, which cause cancer due to steric effect from σ to σ^* transient, which needs a higher energy as the σ bonds are highly stable. Similarly, unsaturated oils do not cause cancer because π to π^* transient needs lower energy as the π bonds are moderately stable. So their quantities are determined along with formulation of cholesterol nanoemulsions based on the theory of like dissolves like.

Experiments

(1) Study the adsorption of acetic acid using the survismeter.

Theory: Under the structure and number theory the number of acetic acid molecules matters a lot but on adsorption by activated charcoal the acetic acid molecules get adsorbed and decrease from the solution. Such adsorption occurs with time and hence with time the number of acetic acid molecules decreases and in the same proportion the friccohesity responds. Thereby the Langmuir theory could be verified using the survismeter. The concept could further be verified with several other acids and adsorbents, like zeolite, calcium silicate, alumina, and titanium oxide.

(2) Determine the salting-out and salting-in effect in water–acetone on adding NaCl salt.

Theory: NaCl salt is added in steps where the water is occupied by the salts and the acetone phase is separated out. In such a phase organic compounds from the crude could be extracted. So it explains the salting out of the acetone phase and then compound in salting in acetone phase from crude product.

(3) Quantitatively determine the absorption of SO_3, SO_2, and CO_2 gases in water, dimethyl sulfoxide (DMSO), and dimethylformamide (DMF) using the survismeter.

Theory: The survismeter detects the molecular populations and hence Millipore water is taken in the RB of the survismeter and friccohesity is determined in a routine manner. Later the selected gas is taken in a container and through a nozzle is connected to the Millipore water filled in the RB. A fixed amount of gas is supplied to the RB water sample and friccohesity is measured. Similarly, again a fixed amount of gas is supplied and friccohesity is measured.

Chapter 56

Determination of the Quality of Soaps and Detergents

Soaps have fatty acids with sodium atoms whereas detergents have sulfate groups, and both groups show different characteristics during the following chemical processes:

- Different nature and level of hydrophilic interactions with R–SO_3Na and R–COONa
- Different effects of R–SO_3Na and R–COONa on the alkyl chain (R is the alkyl chain)
- Different steric effects caused by R–SO_3Na and R–COONa
- Different solvent structure breaking and making activities of R–SO_3Na and R–COONa
- Different Brownian motions of the alkyl chain caused by R–SO_3Na and R–COONa groups

The survismeter along with friccohesity can be used to distinguish soaps from detergents.

Question: Determine the length of the alkyl chain in the following 0.05–2.5 mol of hydrocarbons in methanol for determining intrinsic viscosity:

Survismeter: Fundamentals, Concepts, and Applications
Man Singh and Sunita Singh

ISBN 978-981-4774-70-3 (Hardcover), 978-0-429-02761-1 (eBook)
www.panstanford.com

(i) Pentane
(ii) Hexane
(iii) Heptane
(iv) Octane
(v) Nonane
(vi) Decane

In a nutshell the molecular size and electrostatic points are determined using the following equation:

$$\frac{(\sigma_r - 1)}{n_{\mathrm{mg}}} = B + Dn_{\mathrm{mg}}$$

Here, n_{mg} depicts the number of methylene groups present in the alkyl chains of molecules like surfactants, ionic liquids, and hydrocarbons; σ_r is relative friccohesity; and D is the coefficient. Here, B is noted as the intrinsic friccohesity (IF) and is noted as under:

$$\text{Intrinsic friccohesity} = [\sigma]$$

Further the IF is fitted as under for the determination of hydrocarbons' molecular weights:

$$[\sigma] = kM^a \quad \text{or} \quad \log[\sigma] = \log k + a \log M$$

The IF versus hydrocarbons plot is depicted in Fig. 56.1, where the negative or positive slope depends on the medium nature.

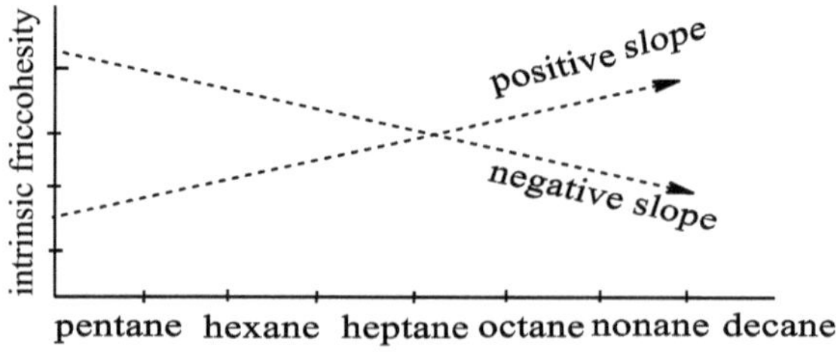

Figure 56.1 Most fabulous science of hydrocarbons with increasing hydrophobicity, which produce friccohesity data within increasing cohesive forces.

56.1 Percolation of Fertilizers

The nanoemulsion form has a higher kinetic energy as compared to the solvation, which has a higher potential energy due to the Lennard-Jones potential (LJP) as the solvent dipoles are attracted by the nuclear charge of the ions. For example, NaCl develops stronger ionic charges on water dipoles and align them around Na^+ and Cl^- nuclei, which have a higher potential energy contrary to nanoemulsions.

The ionic size develops a unique ionic field on developing the ionic hydration sphere. Therefore, increasing ionic sizes weaken the ionic fields in hydration spheres that affect various physicochemical properties during:

- Oil extraction from the crude to the pure state in a solvent
- Ion–dipole interactions that are demonstrated by NaCl and water
- Dipolar–dipolar interactions that are demonstrated by water–dimethyl sulfoxide (DMSO) in a liquid state
- Determination of salt-dissolving capacity of solvent water
- Quantitative determination of saturated oil or fat in an unsaturated oil sample
- Esterification of acetic acid and methanol by selecting their stoichiometry
- Identification of cationic, anionic, and nonionic surfactants using the survismeter
- Determination of the hardness of water by using the survismeter

Theory: The detergents work effectively with pure water because the calcium (Ca^{2+}) and magnesium (Mg^{2+}) cations that are present in hard water consume at least two times the normal amount of detergent. Soap solutions with Ca^{2+} form a soap scum as a white precipitate in place of forming micelles, which trap dust particles. The micelles form lather on trapping air and easily approach the dust particles, contrary to the use of an alkyl chain in forming calcium salt of alkyl chain. The scum is calcium stearate, which is formed from sodium stearate as soap.

$$2C_{17}H_{35}COONa(aq) + Ca^{+}(aq) \rightarrow Ca(C_{17}H_{35}COO)_2 + 2Na^{+}(aq)$$

Similarly, detergents could be explained because they contain sulfate groups in place of carboxylic groups.

$$2C_{17}H_{35} - SO_3Na(aq) + Ca^{+}(aq) \rightarrow Ca(C_{17}H_{35}CO_3)_2 + 2Na^{+}(aq)$$

Hardness is the soap-consuming capacity of a water sample through the precipitation of its alkyl chain. The precipitation prevents lathering of the soap as no micelle formation occurs. Calcium and magnesium carbonates both get deposited as white solids on inside surfaces of pipes and heat exchangers. Overheating of hard water causes calcium carbonate deposition, which blocks the boiler. The deposition varies in crystalline forms like calcite or aragonite. The ions in an electrolyte lead to galvanic corrosion, where one metal is preferentially corroded when in contact with another metal or organic compound when both are in contact with an electrolyte. In swimming pools, water of a turbid or cloudy or milky appearance depicts hard water. Both calcium and magnesium hydroxides are soluble in water and the solubility of hydroxides of calcium and magnesium (alkaline-earth metals) increases moving down the column. Aqueous solutions of these metal hydroxides absorb CO_2 from air, forming insoluble carbonates, giving rise to turbidity. This often results from the pH being excessively high (pH $>$ 7.6) while maintaining a chlorine concentration at the proper level, to lower the pH by adding HCl. Their optimum values are in pH range 7.2 to 7.6.

56.2 Solubility of Alkylbenzenesulfonates Detergent and Soap

***Experiment*:** Determine the solubility of alkylbenzenesulfonate (ABS) detergent and soap (sodium stearate and tristearin, Fig. 56.2) separately in hard water and compare their solubility efficiencies.

***Theory*:** A detergent is a surfactant or a mixture of surfactants with cleaning activities in dilute solutions. The detergents that are ABS are more soluble in hard water as compared to soap. As compared to polar carboxylate (soap) the polar sulfonate strongly binds the

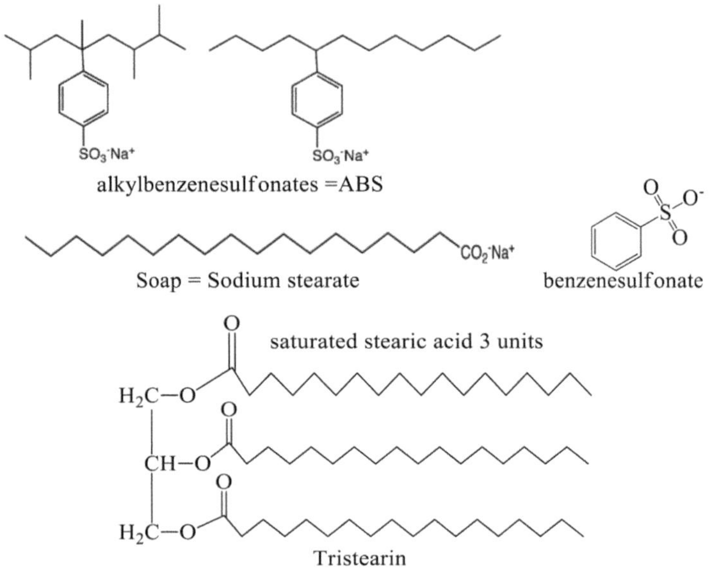

Figure 56.2 Molecules with variable hydrophobicities, which can be branched like ABS and linear like sodium stearate, tristearin, or triglycerides, are used to formulate nanoemulsions.

calcium and other ions found in hard water. The experiments are separately conducted with a prescribed amount of water for the calibration curve of friccohesity versus compositions. The structures are noted as under.

A detergent is a laundry or dish washing agent contrary to a hand soap or other cleaning agents. Detergents are available as powders or concentrated solutions and work like soaps because they are partly hydrophilic (polar) and partly hydrophobic (nonpolar), which is noted as amphiphilic. Their dual nature facilitates the mixture of hydrophobic compounds (oil and grease) with water. Air is not hydrophilic, and the detergents act as foaming agents to varying degrees.

Chapter 57

Hydrophobicity as a Physical Property

Hydrophobicity is a physical property of a hydrophobe like trimesoyl trihexyl malonate ester (TTDHM), which means it does not interact with water, contrary to a hydrophilic substance (Fig. 57.1). Thus TTDHM repels the water and could develop an excellent nanoemulsion. Hydrophobes are nonpolar and prefer other neutral molecules and nonpolar solvents for interacting activities or binding. The dipolar water molecules do not dissolve hydrophobes, so they cluster together, forming micelles. The water on hydrophobic surfaces exhibits a high contact angle. The TTDHM being a stronger hydrophobic material removes oil from water and makes for efficient management of oil spills. It leads to the chemical separation processes to remove nonpolar substances from polar compounds. A hydrophobic material is used interchangeably with lipophilic (fat loving) substances. Hydrophobic substances are lipophilic. Exceptions are silicones and fluorocarbons. Figures 57.1 and 57.2 both highlight the surface areas of crop leaves that are wet. The liquid that completely wets has the least binding energy, high distribution forces (viscosity), and the lowest cohesive forces (CFs). The nonwetting surface also develops an additional air layer, which does not allow more wetting of the crops. Pesticides like endosulfan

Survismeter: Fundamentals, Concepts, and Applications
Man Singh and Sunita Singh

ISBN 978-981-4774-70-3 (Hardcover), 978-0-429-02761-1 (eBook)
www.panstanford.com

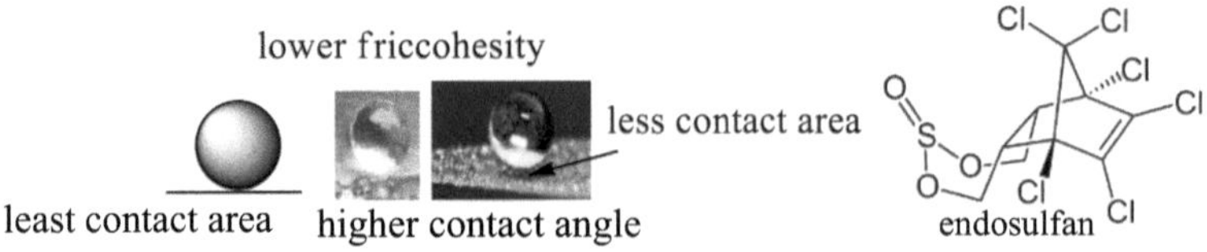

Figure 57.1 Stronger cohesive forces with least contact area of crop leaves with pesticide nanoemulsions and lower friccohesity. Small contact area with higher contact angles form sessile drops of stronger cohesive forces and least adhesive forces. Making nanoemulsions of endosulfan (pesticide) develops higher wettability.

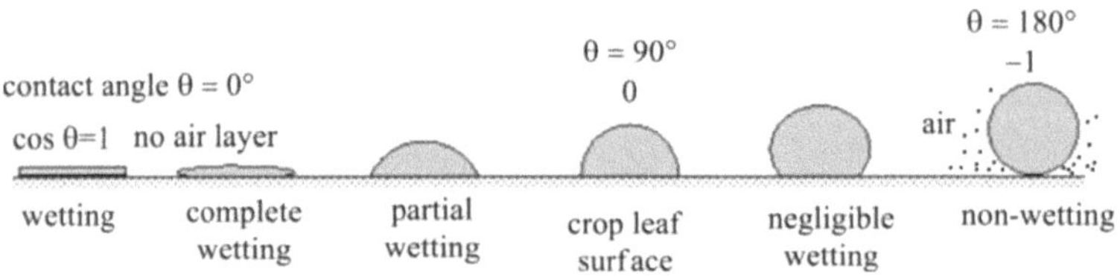

Figure 57.2 Elucidation of wettability and contact angle; higher wettability develops least contact angles. The friccohesity of such nanoemulsions presents an economic model for the use of pesticides and other agrochemicals.

are organic molecules and need more wettability for efficient use as per spreading vis-à-vis the the surface area.

Surface-active agents like biocompatible detergents (like soap) or simple detergents are added in the water, which enhances the wettability of the resultant liquid mixture to be used for agricultural purposes or agripests. Froth formation inhibits coalescence because the hydrophobic localization with a certain amount of loose bonding and the entropic tendencies stabilize the tentropy of atoms and bond oscillations in the molecule. Froth formation in systems of surfactant + water as compared to surfactant + organic solvent indicates that the hydrophobic chains are not much closely packed in structured water and in such loose pockets air enters that expands the hydrophilic thin film at the outer periphery of the micelles. Surfactant + water forms froth, and surfactant + organic solvent forms the mildest froth. Figure 57.3 depicts such distinction of phases since the hydrophobic phase is more structured and is not in

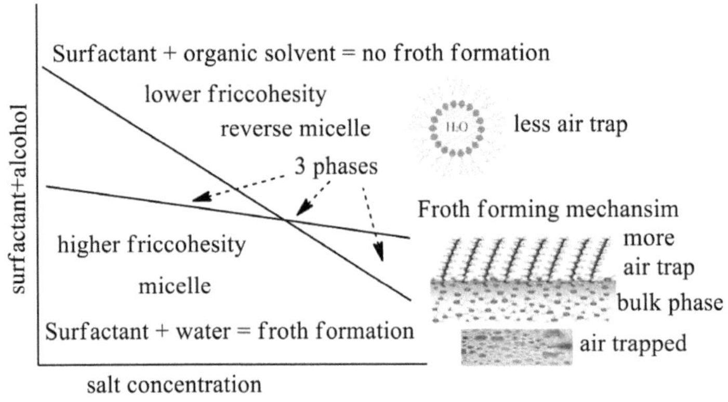

Figure 57.3 Science of friccohesity forming nanoemulsions with froth formation, which traps air inside the micelles. Air pressure expands the thin film of the hydrogen-bonded water placed around the micelles and forms froth or air bubbles.

a position to support air trapping as no gap exists in the hydrophobic chain of surfactants and the hydrophobic part of the oil or the organic phases. However, in the case of the micelles, where water is the main phase, air gaps or gaps where the air is trapped between the multiple hydrophobic alkyl chains are induced.

57.1 Friccohesity and Air Bubble

Trapped air generates pressure and the hydrophilic thin film stretches to support expansion, forming air bubbles of larger sizes; the frictional forces (FFs) and cohesive forces (CFs) play a master role in such processes. These effects and their science and mechanism are covered or explained by friccohesity and are defined as the friccohesive dynamics. The survismeter measurements for interfacial tension (IFT), surface area, particle size, activation energy, tentropy, friccohesity, and intermolecular forces (IMFs) determine the reverse and normal micelle formation. Such sciences lead to the study of the chemical process of exfoliation and functionalization of graphene, which helps develop graphene inks and detective and quantum graphene oxide dots. Fermentation is another useful form of friccohesive dynamics where the binding forces or CFs of the solvent are weakened to develop or form other molecular structures or fragmentations with adequate kinetic energy generated from potential energy. Thus, the friccohesity of fermentation could be determined and hence the nanoemulsion medium is a better medium for a better fermentation with higher kinetic energy and surface areas of the participating molecules. Factors like functional groups, electrostatic poles, lengths of alkyl chains of alcohols, and surfactants participate in formulating nanoemulsions. Therefore, nature of chemical substances influences the microfluidics of nanoemulsions and the wettability of resultant physicochemical properties. Friccohesity critically explains the state of fumigation and floor cleaning detergents because it depicts wettability of the formulations. The highest wettability and spreadibility both are obtained when the CFs are nearly zero and the IMFs are strong. Such mixtures produce the strongest intermolecular interactions with a high viscosity.

Considering the science and mechanism depicted in Fig. 57.3 friccohesity and the survismeter could be considered one of the milestones in physicochemical sciences being applied in the formulation of chemical formulations and their behavior as a medium of heat dissipation.

57.2 Biophysical Applications of Friccohesity

Friccohesity is a fundamental force that monitors and induces adequate circulatory systems through desired boundaries. For example, the circulatory system noted as the cardiovascular or vascular system permits blood to circulate and transport nutrients like amino acids and electrolytes, oxygen, carbon dioxide, hormones, and blood cells to body parts. The cells provide nourishment, stabilize temperature and pH, and maintain homeostasis. Blood flow involves hemodynamics with a definite hemorheology. Such systems work due to friccohesity, where the FFs and CFs work in a most coordinated or tandem manner. For example, higher friccohesity depicts better binding of nutrients with blood for adequate circulation. Conductance is a transport property of salt in the aqueous medium of the body, that is, the blood. The cations and anions move toward the respective electrodes, which generate a current as the size and electron density of the cations and anions are not the same. So they induce a difference in the charges conducted by them. Since the charge is transferred by ions and the ions have different sizes, such properties face different medium resistance. A similar mechanism is also extended to charged molecules, which develop or generate conductance. If the medium is structured due to strong hydrogen bonding (HB) or ionic bonding, conductance is very low. A structured medium or solvent has the highest viscosity and low surface energy or surface tension. High viscosity and low surface tension induce low conductance and high friccohesity. The viscosity is caused by FFs, and the surface tension is caused by CFs. Currently, the product of FFs and CFs is noted as friccohesity. Thus the higher the friccohesity, the lower is the conductance. Such formulations are useful in designing several industrial products, like drugs, inks, and cosmetics.

Chapter 58

Friccohesity and Conductance

New experiments could be developed for finding out the exact relations between friccohesity and conductance with respect to the composition and size of the ions as per the ionic length of the solution. Ionic length (I) is related to friccohesity (σ) as under:

$$\sigma = \sigma^0 + AI$$

Friccohesity directly depends on the size of the particles, so fat or oil molecules have larger droplet sizes because there are stronger cohesive forces (CFs), which hold larger oil amounts together. Thus, the fat is not dispersed in the medium and hence the friccohesity remains lower contrary to the viscosity and surface tension of the aqueous salt solutions, which remain higher than those of the water itself. Therefore, the physicochemical properties like friccohesity is found to depend on hydrophobicity of alkyl chain. The length of alkyl chain containing numbers of $-CH_2-$ groups of surfactant develop hydrophobic interactions with biomolecules like amino acids, proteins, or dendrimers. The number of $-CH_2-$ could be determined by the following equation:

$$\sigma = \sigma^0 + A.n(-\mathrm{CH_2}-)$$

Survismeter: Fundamentals, Concepts, and Applications
Man Singh and Sunita Singh

ISBN 978-981-4774-70-3 (Hardcover), 978-0-429-02761-1 (eBook)
www.panstanford.com

Here, A is the activity of the surfactant series because the hydrophobicity increases for longer chains. Longer chains increase the numbers of the sigma bonds, which strongly hybridize and develop higher potential energy with stronger CFs. Hence, friccohesity could be correlated to hydrophobicity in terms of $-CH_2-$ numbers in the surfactant or the amino acids in the proteins or the tiers in the dendrimers.

$$\sigma = \sigma^0 + A.n(-\mathrm{CH_2}-) \pm B.n(-\mathrm{CH_2}-)^2$$

Here, A is the activity of the surfactant series because the hydrophobicity increases for chain values. Since fluid dynamics is the most continuity state of molecules that are dissolved in liquid state such that their constitutional coordinate orient them to the most stable and optimized liquid state. The NaCl in micelles is trapped and brought in contact with the infected surface, which absorbs the water. Adhesive + NE (nanoemulsion) + NaCl make an efficient mixture. A smart resin like melamine formaldehyde polyvinylpyrrolidone (MFP) could develop an adequate ratio of hydrophilicity to dissolve salts like NaCl and of hydrophobicity to disperse the hydrophobic part to develop a kind of reverse micellar structure. In a rational balance of hydrophilicity and hydrophobicity of this type, the salts could be confined to the core or the central part of micelles, which could have salting in or salting out effects. Also a superhydrophobe like trimesoyl tridihexyl malonate ester (TTDHM) could salt out the water and hydrophilic or solvated ions from one phase to another. The hydrophobic domain with stronger Brownian motions could induce a sound generation and if such motions are amplified then they could be visualized and used as a quality detecting sensor for nanoemulsions. Also a diaphragm could be used for making sound of molecules, which could be placed in the reservoir bulb (RB) sample in the survismeter connected through a Teflon capillary. Capillary science is a natural science that has been supporting human survival by supplying and distribution of oxygen, food, or nutrients. A similar capillary science is responsible for supplying water and minerals in plants. Several sciences work on the basis of capillary theory. In light of such development in capillary sciences several scientists are working and have put forward theories for use.

The survismeter is an expression of the latest wonders of hydrogen bonding (HB) and is practiced in a natural form, that is, put to use as such. Its working does not need external forces like electric, magnetic, and thermal. Capillary action occurs when adhesion to walls is stronger than the CFs between the liquid molecules. The height to which capillary action takes water in a uniform circular tube is limited by surface tension and, of course, gravity. The water sticks together in the form of drops; it sticks to glass, cloth, organic tissues, soil, and, luckily, to the fibers in a paper towel. For example, dip a paper towel into a glass of water and the water climbs up the paper towel. It keeps going up the towel until the pull of gravity is too much for it to overcome.

58.1 Friccohesity and the Survismeter for Fisheries and Pond Life

The survismeter determines the activities of the minerals, gases, and biomass present in the water in ponds. The interacting activities of the minerals, gases, and biomass drastically alter the physicochemical properties of the water and reduce the oxygen absorbing ability of the water in the pond. The stronger CF developing activities of mineral with water prevent the availability of freshwater for fishes. Such activities cause high mortality among aquatic life and lead to a great loss in the fisheries and the agriculture industries. The survismeter works on the basis of fluid dynamics where molecular interactions, ionic interactions, and salting binding abilities occur frequently. Such physicochemical activities generate specific surface tension, interfacial tension (IFT), particle size, viscosity (distribution), friccohesity, activation energy, and cationic and anionic solvation spheres and engage the water by affecting the normal activities of the water and ultimately lead to the deterioration of pond life. The survismeter measures these parameters together without any need for external force like electricity and magnetic forces. It is easy to use by farmers in the fishery industries.

58.2 Molar Volume and Molar Friccohesity as Interacting Constituents

The survismeter is a green labtech (laboratory technology) mosaic of molecular sciences reflecting the intriguing absolute mechanism of the chemical process driven by physicochemical forces caused by the electronic configuration of molecules. There are two fundamental steps. Number one is to prepare the nanoemulsion, and number two is to subject the nanoemulsion to the survismeter. The survismeter is the characterizing science that works based on the actual fixed boundary condition in 1D capillary science to exactly reflect the chemical linkage of interacting constituents of the mixture or the mixture forming constituents. The chemical activities with a definite dimension that develop the geometrical shapes of the materials are studied with fluid dynamics materialized or conducted with the survismeter. The survismeter allows differential molecular activities in continuum as well as in noncontinuum modes. The molecules have atoms, shared electrons, unshared electrons, and shifted electron pairs, which express their language in terms of developing a specific order of friccohesity of fluid dynamics vis-à-vis survismeter science. The molar volume depends on the molecular interaction engineering, which is initiated by the overall structural frameworks of the molecules. This engineering produces molar volume and friccohesity of definite magnitudes. Thus, both molar volume and friccohesity matter a lot for correlating molar volume with friccohesity. Both molar volume and molar friccohesity depend on the interacting constituents of the molecules either with similar molecules that develop molar volume and friccohesity of pure molecules or with dissimilar molecules. Such liquid mixtures develop specific physicochemical parameters that determine their interacting activities. These parameters are most critical in facilitating drug designing, their binding into a complex, and their release with time after reaching the wanted cell site.

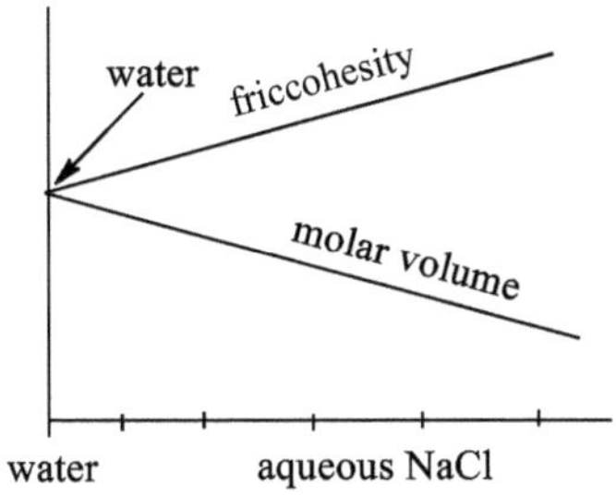

Figure 58.1 A unique model for representing the friccohesity and molar volumes of aqueous NaCl solutions.

Similarly, as the molar volume directly depends on the density, the relationship of the density and friccohesity could be most logical to explain the molecular interactions. In general, in the case of the pure samples the stronger the CFs or van der Waals forces, the lower is the molar volume and the friccohesity. However, the pattern trend for liquid mixtures depends on the strength of the interactions where for stronger solute and solvent interactions, the molar volume is lower but the friccohesity becomes higher.

Question: Determine the strength of the solute and solvent interactions by preparing 0.05, 0.10, 0.15, and 0.20 mol kg^{-1} aqueous NaCl solutions and measuring their molar volumes and molar friccohesity (Fig. 58.1).

58.3 Friccohesity as a Unique Science

Molecular constituents establish the overall framework of an individual molecule and also establish the interacting abilities of the molecule with itself or with other molecules. The electronic orbitals are obtained or developed as per the distribution patterns of the electrons, for example, the highest occupied molecular orbital (HOMO) and the lowest unoccupied molecular orbital (LUMO), which are distributed in 3D (Fig. 58.2).

Electronic transitions could be attained by mechanical energy, photon absorption, heating, and the magnetic field, and such variations could be applied to the solution sample in the RB of

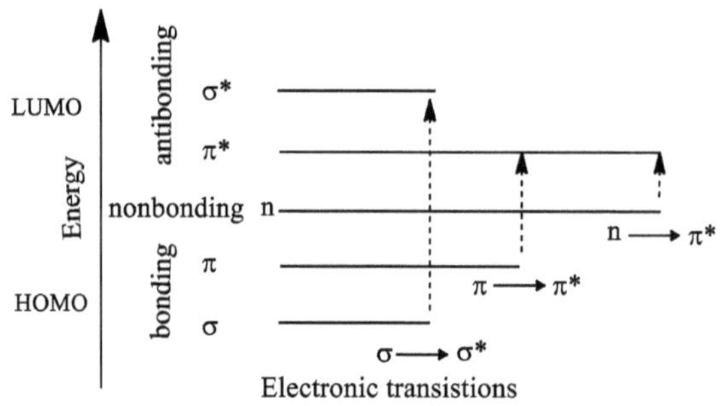

Figure 58.2 Electronic transitions of lone pair and π and σ bonds electrons through HOMO to LUMO orbital models with their potential energies.

the survismeter. Thereby the survismeter seems to be a most competent device to track the electronic changes in the LUMO and HOMO energy profile by developing unique fluid dynamics and friccohesity and tentropy values. These electronic transitions occur in a single molecule where the molecule itself gains stress and strain under a rheological aptitude. These changes in single molecules are explained by the tentropy and intramolecular multiple forces theory (IMMFT). Also the ionic hydration or the solvation sphere is developed in 3D in the nanometer area. Constructively, the fluid flow of such liquids in a 1D capillary streamlines linearly or in laminar order, which facilitates the measurement of viscosity. Thus viscosity is exemplified in 1D due to flow in the 1D capillary of a fixed diameter; even if the fluid flow is non-Newtonian, the changes remain in 1D only. Contrary to the viscosity the fluid dynamics of the pdn in the case of surface tension give freedom to express the liquid spreading in 3D similar to the electronic distribution in 3D. Though electronic distribution is the most prominent and integrated process of fluid flow, it cannot be denied that surface tension or surface energy gives an opportunity for molecular forces to spread in 3D similar to an expression electronic orbital in 3D. Fundamentally, the viscous flow with definite forces and the pdn with another set of molecular forces (adhesive forces) are the main constituents or components of friccohesity. Thus friccohesity is the most advanced version of the expression of molecular forces incorporating electronic sciences (Fig. 58.3).

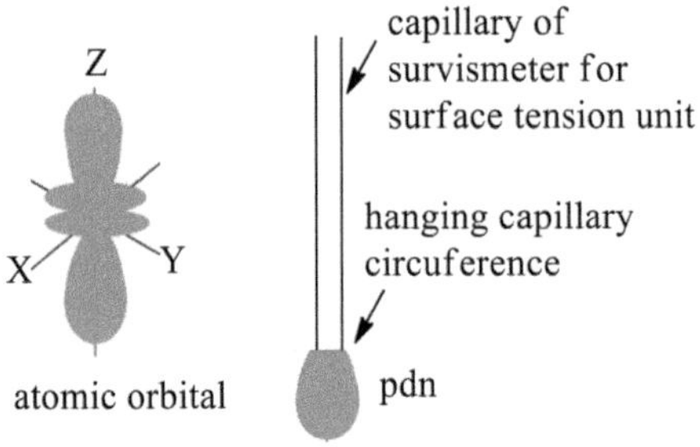

Figure 58.3 Dimension of atomic orbital modeled to extract out a working science of pendant drop formation at the bottom tip of the hanging capillary in CPU of the survismeter.

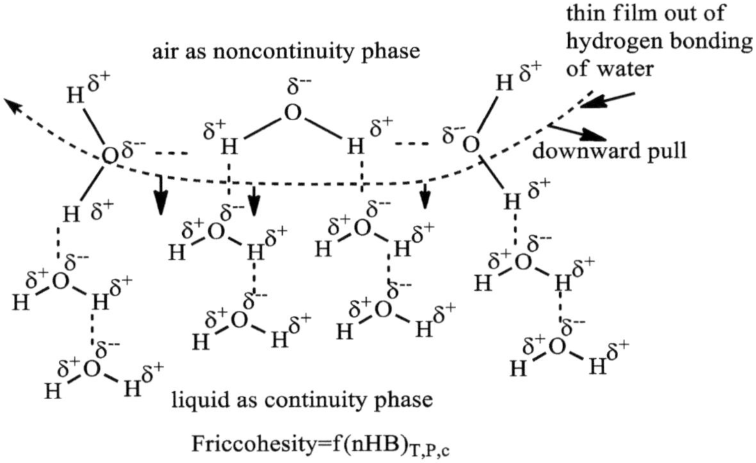

Figure 58.4 Air–liquid interface of water, which forms a thin film on sharing its residual forces as vectors. These unbalanced forces at air–liquid interfaces develop surface tension. Contrary to the mechanism of residual forces of molecules generating surface tension, these forces in the bulk phase are counterbalanced by force vectors of the neighboring or nearby molecules.

Friccohesity is inferred from the ability of stretched interhydrophilics of the head group of surfactant, which is depicted in Fig. 58.4. The air–water forms air bubbles out of the bionanoemulsion as the air enters its micelles.

Chapter 59

Thermodynamics and Kinetic Stability Mechanism of Micelles

Oil and water nanoemulsions with (cationic, anionic, and nonionic) surfactants vis-à-vis their thermodynamic and kinetic stabilities could be elucidated by the study of equivalence of hydrophilic to hydrophilic and of hydrophobic to hydrophobic, respectively. For normal micelle formation the oil is stoichiometrically dispersed to the core of micelles through the hydrophobic alky chain as the tail of the surfactants. Thus the ratio of both surfactants and oil critically develops the stability of the micelles. Similarly, hydrophilic water acts as a bit buffer zone contrary to the hydrophobic components because at the periphery the hydrophilic head engages the water through hydrogen bonding (HB). The water develops a continuous thin film at the air–liquid interface (ALI), which remains tense as the pulling forces work downward to balance the overall energy of the liquid. This is the only reason that the pulling forces are noted as the surface energy of a tense thin film. The intensity of pulling depends on the strength of the molecular forces of the liquid. A single oil molecule holds three times the length of alkyl chain of fatty acids that a single molecule of a surfactant holds. The surfactant has one alkyl chain attached with the one head group. Therefore, at least

Survismeter: Fundamentals, Concepts, and Applications
Man Singh and Sunita Singh

ISBN 978-981-4774-70-3 (Hardcover), 978-0-429-02761-1 (eBook)
www.panstanford.com

one-third the amount of oil is always used as compared to surfactants. Such fascinating surface science could not be treated in isolation of the bulk because only the molecular forces of the bulk phase that are near ALIs are responsible for pulling as the electrostatic forces cause stronger molecular forces, which are responsible for a stronger pull or higher surface tension values.

59.1 Extraordinary Experiments Conducted with the Survismeter

The survismeter measures physicochemical properties (PCPs) of liquid mixtures, which act as an indicator for the quality or stoichiometric ratios. Some of the PCPs are as follows:

- Surface tension: Default surface science domain
- Dynamic surface tension: Air–water bubble (AWB) stretching cohesive forces (CFs)
- Kinematic surface tension: Capillary rise activity
- Static surface tension: Pendant or sessile drop
- Liquid–liquid interfacial tension (IFT): Friccohesity on 10–100 KJ/mol scale
- Biocompatible food membrane: Dynamic membranes
- Color-changing experiment of micelles versus $KMnO_4$
- Wonders of color changes in reverse micelles versus ink or phenolphthalein indicator
- Wettability: CF model to centrifugal force
- Viscosity: Chemical interweave of constituents in the solvent phase
- Surface area: Strength of the O–H (secondary) bond with potential energy
- Particle size: Interhydrophobics and interhydrophilics (inter-hydrodynamics)
- Interodynamics: AWB continuity for the pseudoboundary model
- Particle size of a nanoemulsion due to interfacing
- Activation energy: Out of disruption of HB
- Structural energy for chemical activity

- Oil and water phases: Activation energy zero
- Glucose water activation: $\Delta E < 0$
- Oil and water inactive phase to nanoemulsion as the active phase
- HB as the source of activation energy
- Friccohesity for molecular interaction engineering
- AWB out of a bionanoemulsion: Stretching ability of the head group
- Wonders and ability of hydrophilics: Friccohesity
- Bionanoemulsion: Medical wonder
- Bionanoemulsion for air, oxygen, hydrogen, and CO_2 bubbles
- Bottled nanoemulsion fitted with an inlet for gas intake
- Tooth versus AWB for quality check of the toothpaste
- Soap-detergents versus AWB for quality check
- Wettability of ink: CF dynamics
- Wettability and monodisperse science of nail polish
- Molecule-to-molecule linkages: Ideal nanoemulsions
- Molecular rheology for thermodynamics stability
- Rheology and tentropy of molecules: Asymmetric involvement of molecules in chemical processes
- Heating molecules and tentropy: Intramolecular asymmetric changes
- Normal tentropy of crystalline molecules with a definite lattice
- Molecular activity vis-à-vis solvent: Electrostatic poles
- Webdrimer and drug encapsulation
- Cell functioning and friccohesity
- Cell functioning with respect to the friccohesity of cell culture of cell fluid
- Friccohesity of normal cell and cancerous cells
- Interaction of calcium oxalate with blood cell
- Strength of nano thin films of water, dimethyl sulfoxide (DMSO), dimethylformamide (DMF), and acetonitrile by measuring their surface tension vis-a-vis their dipole moment with the survismeter
- Quality of glucose water being administered to the patient in an emergency

Note: The viscosity of blood is the most sensitive physicochemical sensor to observe the supply of oxygen to organs. If the blood viscosity is higher then it costs much stress and strain vis-à-vis arteries because higher viscosity causes stronger shear stress on the arteries. Similarly, lower viscosity with lower shear cannot hold and absorb adequate amount of oxygen and therefore supplies less amount of the oxygen to the blood.

59.2 Determine Wettability of Pesticides

- Determine the wettability and accumulation of endosulfan or dieldrin or aldrin in its aqueous solution using the survismeter.
- Determine the thermodynamics and kinetic stability of the endosulfan aqueous solution.

Note: When farmers prepare aqueous solutions of pesticides, for the time being the formulations remain homogenized but with time coagulation starts and the pesticides cannot be taken by the root membrane due to size and higher potential energy. So such determinations are necessary.

- Determine the efficiency of dish-washing liquids using the survismeter.

Note: Prepare their aqueous solutions and determine the friccohesity where the higher value of friccohesity predicts higher efficiency of the material.

Chapter 60

Measurements of Physicochemical Properties with the Survismeter

Caution: The survismeter is a delicate laboratory device. It needs care in its handling and for doing experimental measurements with.

60.1 Sample Filling in the Survismeter

First Step: Sample Filling in the RB of the Survismeter

To begin with, for the sample filling, take the survismeter unit out of the box made of the transparent polyacrylic sheet used as a water bath. Carefully hold the survismeter unit in your hands. The reservoir limb is comparatively robust, so hold it at a 45° angle. This spatial position of the survismeter prevents micro air bubbles from getting trapped inside the reservoir bulb (RB) during sample filling. When the survismeter is not in use naturally, air enters its internal empty spaces. On sample filling this air comes out of the survismeter's empty spaces. Carefully hold the survismeter unit with left hand and hold the sample vessel/container with the right hand. Ensure you are carefully holding the survismeter and the sample as well. Ensure that the survismeter unit is inclined

Survismeter: Fundamentals, Concepts, and Applications
Man Singh and Sunita Singh

ISBN 978-981-4774-70-3 (Hardcover), 978-0-429-02761-1 (eBook)
www.panstanford.com

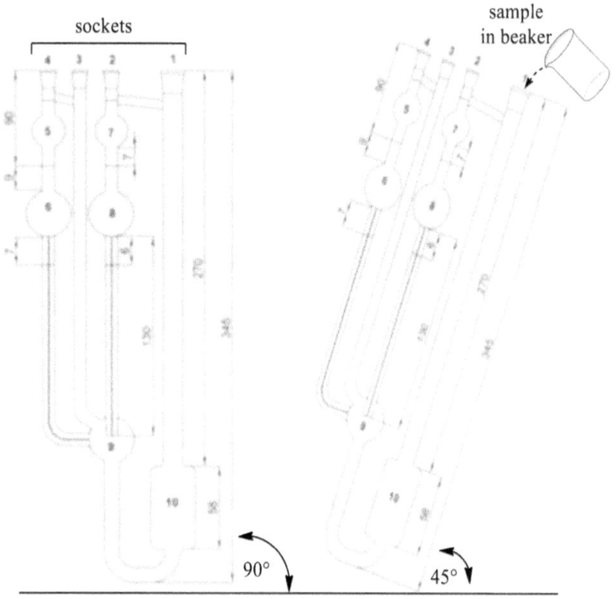

Figure 60.1 Spatial impact on the shear stress of the liquid samples for measuring microfluidics. Such positional configuration of survismeter determines Brownian motions including other molecular motions like notational, translational, vibrational.

at a 45° angle with respect to the horizontal or earth plane. The positions are shown in Fig. 60.1. The nose of the sample holder is carefully brought in close contact of the socket at the top of the reservoir limb (Fig. 60.1). However, the volatile and surfactant liquid samples can be transferred through a silicon tube fitted to the extruded part of the syringe (Fig. 60.2). After filling the sample in the RB fix the stoppers in the sockets except one—either the surface tension or the viscosity unit. For example, first the surface tension is measured. Then the cone that is attached with the silicon tube is fixed with the surface tension unit and stoppers are fixed in the sockets of the pressure limb and viscosity units. The stopper is not fitted in the socket of the reservoir limb because this unit allows the head-on pressure needed for taking up the liquid sample to the surface tension and viscosity units. Therefore, the socket

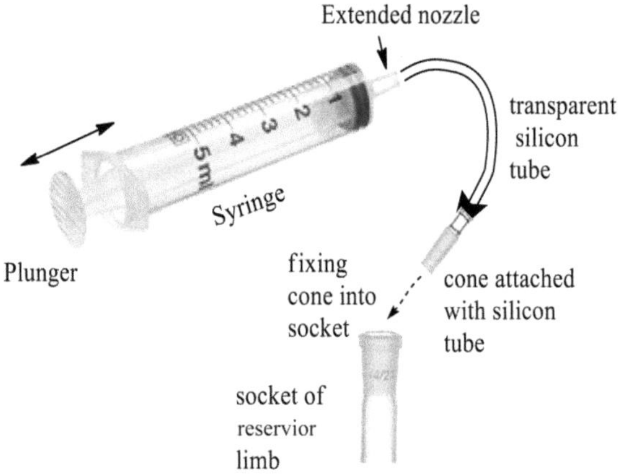

Figure 60.2 Syringe and plunger attached to hollow glass cone through a transparent soft silicon tube acts as a piston to regulate and to evacuate the pressure from the capillaries of the survismeter units. The cone is attached to the socket of the survismeter units at the top of the survismeter.

and the cone are the most significant joint venture for successful operation and handling of the survismeter. The socket-and-cone arrangement is noted as a local interprocess communication (IPC) with the survismeter. So care should be taken to attend the IPC without any air passage. The socket and cones allow communication of the survismeter with the sample that is under measurement.

Second Step: Sample Filling in Functional Units

There are many steps of sample filling, but the major steps are two. Initially, the sample is transferred from the beaker or vessel to the RB of the survismeter. Secondly, the sample filled in the RB is lifted to the working units. These working units are surface tension, viscosity, and others.

60.2 Surface Tension Measurements

For measuring surface tension the liquid sample is lifted from the RB to bulb 8 and to two-thirds of bulb number 7. During this operation sockets number 3 and 4 are stopped so that the sample cannot go to them and the sample is pressure-guided to move to the surface tension unit. For this purpose the cone of the piston pump or the syringe is fitted in the socket of the surface tension unit. It should be airtight, and the pressure in the surface tension unit is reduced, which is compensated by the liquid sample. When two-thirds of bulb number 7 is filled then stop the pressure reducing operation and carefully withdraw the cone from the surface tension socket, followed by removing the stopper from the socket of the pressure limb. Pendant drop numbers/counts (pdns) being developed at the circumference of the hanging capillary inside the controlled pressure unit (CPU) are counted with the pdn counter. The pdns are counted out of the liquid volume filled in bulb 8 within its upper and lower marks. The measurements are made for solvents as well as for sample.

60.3 Viscosity Measurements

A similar procedure is applied for viscosity measurements except that the cone of the syringe is fixed with the viscosity socket.

60.4 Description of Survismeter Parts

Numbers 1, 2, 3, and 4 depict sockets of their respective units. Numbers 5 and 6 are the buffer bulb (BB) and the functional bulb (FB) of the viscosity unit. Numbers 7 and 8 are similar bulbs of the surface tension unit. Socket number 3 belongs to the pressure limb, which is connected with the CPU. The pressure limb equilibrates the pressure of the CPU unit for filling up the sample in the respective unit and also during the measurement of the data of surface tension, viscosity, and others. Numbers 9 and 10 show the CPU and the RB, respectively.

Now slowly fill the RB to the mark by allowing the sample to flow from a vessel/beaker just touching the reservoir limb. Sample filling in this way reduces the chances of air contamination and trapping of air due to froth formation in case the sample solution is of a surfactant or a protein. After properly filling the sample, place the survismeter unit inside the polyacrylic water bath and place the Sachin plate over the unit to hold it at a 90° angle.

60.4.1 Multicast Sockets

These sockets are significant for establishing a connection between the survismeter fluid dynamics of the liquid samples. Filling a sample in the RB is the first step to connecting the survismeter with the liquid sample, but the sockets play a significant role in taking the sample from P2P (point to point). The samples can be usefully lifted to the surface tension unit (STU) or the viscosity unit (VU) via sockets vis-à-vis their respective virtual valves opening into the CPU.

60.5 Measurement of Surface Tension (01:10 to 06:18 min.)

After taking the sample in the RB, close the pressure limb and the viscosity limb with the operational stoppers. Place the L-shaped glass adaptor of the syringe pump over the surface tension limb and close it tightly, making it airtight. Now pull the plunger of the syringe pump slowly and you will see the sample getting into the CPU. Continue pulling the plunger and you will observe that the sample starts to fill into the surface tension unit. Pulling the plunger slowly helps in maintaining homogeneous pressure inside the survismeter unit and maintains uniform pressure distribution among the constituents of the liquid sample. The surface tension limb contains two bulbs: the lower bulb is the FB, and the upper one is the BB. Fill the surface tension limb till the BB is half filled. The role of the BB is to give the user extra time to start the analysis. This also allows the constituents of the sample to achieve equilibrium since extra pressure has been used to lift the sample in the surface tension limb, increasing the overall kinetic energy of the system. This matters a lot when the sample is a protein or surfactant solution, and this buffer zone helps the constituents to attain the same energy state where they were before the start of the analysis. After equilibrium is achieved, open the glass adaptor of the syringe from the surface tension limb. There will be no pendant drop formation in the CPU. To start the drop formation, open the operational stopper from the pressure limb. Drop formation occurs because the flow of the sample is stopped midway and the sample falls in the form of a drop when the weight of the drop crosses the tension on the air–liquid interface (ALI) of the sample. When the sample solution crosses an upper mark located between the FB and the BB of the surface tension limb, start counting the number of pendant drops formed using the pendant drop counter. Stop counting the drops once the sample crosses the mark located below the FB. Repeat the experiment without taking the sample out from the RB in order to find out the repeatability associated with the experiment. For surface tension calculation, in addition to the number of drops formed for the sample solution between the upper and lower mark, the number of drops formed between the upper and lower marks of the surface tension limb for a reference liquid is also required at the experimental temperature. The solvent in which the sample solution is prepared is generally selected as the reference liquid. For the surface tension limb, at a given temperature, the number of pendant drops formed for a particular reference liquid will always be constant and hence with every sample there is no need to find the number of pendant drops formed for the reference liquid as it can be taken from the previous experiment at that temperature. For example, suppose the experimental temperature is 25° Celsius and the sample is 0.1% weight by weight of the aqueous solution of any water-soluble protein. The reference liquid will be water. The number of pendant drops formed for water can be taken from any previous experiments where the number of drops was counted for water at 25° Celsius. Now, to calculate the surface tension of the sample, the pdns of the sample and the reference liquid are used in the formula as follows:

Calibration Constant of Surface Tension

$$k_{\mathrm{ST}} = \frac{\gamma_0 n_0}{\rho_0}$$

Surface Tension Calculation

$$\gamma = \left(\frac{n_0}{n}\right)\left(\frac{\rho}{\rho_0}\right)\gamma_0$$

where,
n_0 = PDN of reference liquid
n = PDN of sample
ρ_0 = Density of reference liquid
ρ = Density of sample
γ_0 = Surface tension of reference liquid
γ = Surface tension of sample

Let reference liquid is acetonitrile and sample in methanol

$n_0 = 283$, $n = 365$, $\rho_0 = 0.778301\ \text{g.cm}^{-3}$, $\rho = 0.788749\ \text{g.cm}^{-3}$, $\gamma_0 = 28.20\ \text{nNm}^{-1}$, $\gamma = ?$

Since

$$k_{\text{ST}} = \frac{\gamma_0 n_0}{\rho_0}$$
$$k_{\text{ST}} = \frac{28.20 * 283}{0.778301}$$
$$k_{\text{ST}} = 10253.87$$

Since

$$\gamma = k_{\text{ST}} * \rho / n$$

Hence,

$$\gamma_{\text{methanol}} = 10253.87 * 0.788749/365 = 22.16\ \text{mNm}^{-1}$$

Literature value for

$$\gamma_{\text{methanol}} = 22.10\ \text{mNm}^{-1}$$

Therefore,

$$\Delta\gamma = 0.06\ \text{mNm}^{-1}$$

60.6 Measurement of Viscosity

After taking the sample in the RB, close the pressure limb and the surface tension limb with the operational stoppers. Place the glass adaptor of the syringe pump over the viscosity limb and close it tightly, making it airtight. Now pull the plunger of the syringe pump slowly and you will see the sample getting into the CPU. Continue pulling the plunger and you will observe that the sample starts to fill into the viscosity unit without going to the surface tension limb. Pulling the plunger slowly helps in maintaining homogeneous pressure inside the survismeter unit and maintains uniform pressure distribution among the constituents of the liquid sample. The viscosity limb contains two bulbs: the lower bulb is the FB, and the upper one is the BB. Fill the viscosity limb until the BB is half filled. The role of the BB is to give the user extra time to start the analysis. This also allows the constituents of the sample to achieve proper orientation since extra pressure has been used to lift the sample in the viscosity limb, increasing the overall kinetic energy of the system. This matters a lot when the sample is a protein or surfactant solution, and this buffer zone helps the constituents to attain the same energy state with ordered orientation where they were before the start of the analysis. After equilibrium is achieved, open the glass adaptor of the syringe from the viscosity limb. There will be no formation of viscous flow seen in the CPU till now. To start viscous flow formation, open the operational stopper from the pressure limb. The viscous flow formation is continuous because the flow of the sample is not stopped midway and the sample solution continuously flows due to gravity. When the sample solution crosses the upper mark located between the FB and the BB of the viscosity limb, start the digital timer and stop the timer once the sample crosses the mark located below the FB and record the viscous flow time taken between the two marks. Repeat the experiment without taking out the sample from the RB to find out the repeatability associated with the experiment. For viscosity calculation, in addition to the viscous flow time recorded for the sample solution, the viscous flow time recorded between the two marks of the viscosity limb for a reference liquid is also required at the experimental temperature. The solvent in which the sample solution is prepared is generally selected as the reference liquid. For the viscosity limb, at a given temperature, the viscous flow time for a particular reference liquid will always be constant and hence with every sample, there is no need to find out the viscous flow time for the reference liquid, as it can be taken from the previous experiment at that temperature. For example, suppose the experimental temperature is 25° Celsius and the sample is 0.1% weight by weight of the aqueous solution of any water-soluble protein. The reference liquid will be water. The viscous flow time for water can be taken from any previous experiments where the viscous flow time was recorded for water at 25° Celsius. Now, to calculate the viscosity of the sample, the viscous flow times of the sample and the reference liquid are used in the formula as follows:

Calibration Constant of Viscosity

$$k_{\text{vis}} = \frac{\eta_0}{t_0 \rho_0}$$

Viscosity Calculation

$$\eta = \left(\frac{t}{t_0}\right)\left(\frac{\rho}{\rho_0}\right)\eta_0$$

where,
t_0 = VFT of reference liquid
t = VFT of sample
ρ_0 = Density of reference liquid
ρ = Density of sample
η_0 = Viscosity of reference liquid
η = Viscosity of sample

Let reference liquid is acetonitrile and sample in methanol

$$t_0 = 128.02,\ t = 194.78,\ \rho_0 = 0.778301\ \text{g.cm}^{-3},$$
$$\rho = 0.788749\ \text{g.cm}^{-3},\ \eta_0 = 0.3430\ \text{cP},\ \eta = ?$$

Since

$$k_{\text{vis}} = \frac{\eta_0}{t_0 \rho_0}$$
$$k_{\text{vis}} = \frac{0.343}{128.02 * 0.778301}$$
$$k_{\text{vis}} = 0.0034425$$

Since

$$\eta = k_{\text{ST}} * \rho * t$$

Hence,

$$\eta_{\text{methanol}} = 0.0034425 * 0.788749 * 194.78$$
$$= 0.5288\ \text{cP}$$

Literature value for

$$\eta_{\text{methanol}} = 0.5440 \text{ cP}$$

Therefore,

$$\Delta\eta = 0.0152 \text{ cP}$$

60.7 Measurement of Interfacial Tension

Clean and absolutely dry the interfacial tension (IFT) capillary so that uninterrupted capillary action can occur. In a clean beaker take the high-density liquid (HDL) and place the IFT capillary in it. Put the glass adaptor of the syringe over the IFT capillary and pull the plunger. Fill the IFT capillary to half of the BB. Once the BB is half filled, take out the glass adaptor and put the operational stopper over the IFT capillary to make it airtight. Now fill the RB with the low-density liquid (LDL) up to the level where the IFT capillary just touches its surface. Once this is done, place the IFT unit over the reservoir limb and allow the systems to reach the equilibrium temperature in which the study is being conducted inside the polyacrylic water bath. Once equilibrium is achieved, open the operational stopper and count the number of pendant drops formed for the HDL in the LDL between the upper and the lower mark. Once the pendant drops have been counted, take out the sample from the RB and again count the number of pendant drops formed for the HDL, but now in air and not in the LDL. Now, to calculate the IFT between the two immiscible liquids, the number of pendant drops of HDL formed in air and in the LDL is used in the formula as follows:

$$\gamma_{\text{IFT}} = \left(\frac{n_{\text{HDL in air}}}{\text{n}_{\text{HDL in LDL}}}\right)\left(\frac{\rho_{\text{HDL}} - \rho_{\text{LDL}}}{\rho_{\text{HDL}}}\right)\gamma_{\text{HDL}} \qquad (60.1)$$

Here, n_{HDL} and n_{LDL} are pdns of the HDL and LDL, respectively; ρ_{HDL} and ρ_{LDL} are densities of the HDL and LDL, respectively; and γ_{HDL} is the surface tension of the HDL.

60.8 Measurement of Contact Angle (θ) and Wetting Coefficient

Clean and absolutely dry the IFT capillary so that uninterrupted capillary action can occur. In the RB, take the sample whose contact angle and wetting coefficient is to be determined. The sample solution should be filled up to a level where the IFT capillary just touches the sample solution. Now, take the cleaned and dried IFT capillary and make it airtight with the operational stopper and place it inside the reservoir limb, allowing the system to reach the equilibrium temperature at which the study is to be conducted. Remove the operational stopper and start the digital timer. On removing the operational stopper, you will see the sample solution starts to rise inside the IFT capillary due to the capillary action. Stop the digital timer once the rise in capillary stops. Note down the time of capillary rise and the height up to which the sample solution rises in the IFT capillary. To measure the wetting coefficient, prepare sample solutions of different concentrations and note down the time of capillary rise and the height up to which the sample solution rises in the IFT capillary. During every capillary rise analysis, always clean and dry the IFT capillary for precise data. Now, to calculate the contact angle and the wetting coefficient (w_c^0), the time of capillary rise and the height are used in the formula as follows:

$$\frac{\eta h^2}{t} = w_c^0 + A\gamma + B\gamma^2 \quad (60.2)$$

The wetting coefficient is determined by determining the sample's upward column rise or height h in time t sec. of the liquid in the IFT capillary. The capillary is absolutely dried before taking the reading, and it is ensured that the inner diameter of the capillary is uniform through its length. The h is column rise in time t, A and B are coefficients, γ is the surface tension, and η is the viscosity of the sample. The $(\eta h^2/t)$ versus γ plot gives the wetting coefficient value on extrapolating the graph to zero compositions.

The contact angle is calculated using the column rise height h and surface tension data (γ, mN/m); ρ is the density, g the gravitational force, and r the radius of the IFT capillary.

$$\frac{h\rho g r}{2\gamma} = \cos\theta \quad (60.3)$$

On putting $r = 0.05$ cm, column rise $h = 2.5$ cm, $g = 980$ cm s^{-2}, and surface tension $\gamma = 72.8$ mN/m in Eq. 60.3, the contact angle θ for water used as sample at 20° C temperature:

$$\cos\theta = \frac{2.5\ \text{cm} \times 0.99821\ \text{g cm}^{-3} \times 980\ \text{cm s}^{-2} \times 0.05\ \text{cm}}{2 \times 72.8\ \text{g cm s}^{-2}\text{cm}^{-1}}$$

The θ value is found to be $\cos\theta = 0.8398$ or 0.84 or $84/100 = 21/25, \theta = 0.99989$ or $\theta = 1$.

Chapter 61

Mansingh Constant (M_c)

Friccohesity (σ, sec. cm^{-1}) is a product of cohesive forces (CFs) and frictional forces (FFs) that result from preparing liquid mixtures. It is determined using the newly formulated equation

$$\sigma = \sigma_0 \left[\left(\frac{t}{t_0} \pm \frac{B}{t} \right) \left(\frac{n}{n_0} \pm 0.0012(1-\rho) \right) \right] \tag{61.1}$$

Here σ_0 is the friccohesity of the solvent, t and t_0 and n and n_0 are as usual, B is the capillary constant and B/t is kinetic energy correction, and ρ is the density of the sample. Aqueous melamine-formaldehyde-polyvinylpyrrolidone resin solutions were prepared for measuring their friccohesity and viscosity, plotted in Figs. 61.1 and 61.2. In the case of highly reproducible and accurate equipment the values of B/t and $0.0012(1-\rho)$ (buoyancy correction) range from 10^{-7} to 10^{-6} and are omitted when Eq. 61.1 becomes Eq. 61.2.

$$\sigma = \sigma_0 \left[\left(\frac{t}{t_0} \right) \left(\frac{n}{n_0} \right) \right] \tag{61.2}$$

Equation 61.2 can be rearranged as Eq. 61.3 by taking the parameters of the solvent or the reference liquid sample out of the brackets, depicted as under:

$$\sigma = \sigma_0 \left[\left(\frac{tn}{t_0 n_0} \right) \right] \tag{61.3}$$

Survismeter: Fundamentals, Concepts, and Applications
Man Singh and Sunita Singh

ISBN 978-981-4774-70-3 (Hardcover), 978-0-429-02761-1 (eBook)
www.panstanford.com

Further Eq. 61.3 is noted as under:

$$\sigma = \frac{\sigma_0}{t_0 n_0}\,[(tn)] \tag{61.4}$$

The term depicted in Eq. 61.5 is named Mansingh constant.

$$\frac{\sigma_0}{t_0 n_0} = M_c \tag{61.5}$$

Equation 61.5 is depicted as Eq. 61.6.

$$\sigma = M_c\,[(tn)] \tag{61.6}$$

M_c is Mansingh constant. It is a product of t as viscous flow time and n as the CF and shows direct relations as under:

$$\sigma \alpha [(tn)] \tag{61.7}$$

Thereby, friccohesity at a constant temperature (T), pressure (P), and the mole fraction of solute n_1 and of solvent n_2 could be expressed as under:

$$\text{Friccohesity, } \sigma = M_c (tn)_{T,P,n_1,n_2} \tag{61.8}$$

Similarly, the friccohesity of mixtures is formulated as under within dT and dP, which are variations in temperature (T) and pressure (P) for mole fractions of solute n_1 and of solvent n_2.

$$d\sigma_{\text{mix}} = \left(\frac{\partial \sigma}{\partial T}\right)_{T,n_1,n_2} dT + \left(\frac{\partial \sigma}{\partial P}\right)_{p,n_1,n_2} dP + \left(\frac{\partial \sigma}{\partial n_1}\right)_{T,P,n_2} dn_1 + \left(\frac{\partial \sigma}{\partial n_2}\right)_{T,P,n_1} dn_2 \tag{61.9}$$

The variations in friccohesity within the limits from 1 to 2 could be noted as Eq. 61.10.

$$d\sigma_{\text{mix}} = \sum_{1}^{2} \sigma_i n_i \tag{61.10}$$

For $dT = 0$ and $dP = 0$, the friccohesity of mixtures could be expressed as Eq. 61.11.

$$\frac{\sigma}{M_c} = [t_A n_A x_A + t_B n_B x_B + t_C n_C x_C] \tag{61.11}$$

Figure 61.1 finds clearly resolved lines without any overlapping, but Fig. 61.2 finds overlapped points on the curves, which infers that when a product of the CFs and FFs are considered then it gives a

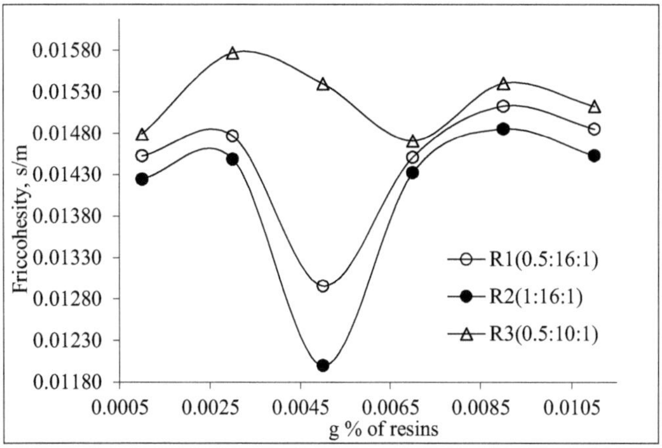

Figure 61.1 Friccohesity of melamine–formaldehyde–polyvinylpyrrolidone (MFP) polymer resin with its increasing percentage composition in its (MFP) aqueous solutions. The R_1, R_2, and R_3 represent variable stoichiometric compositions of melamine, formaldehyde, and polyvinylpyrrolidone in MFP resins, respectively. No overlapping of friccohesity data for any of the three resin stoichiometries is observed.

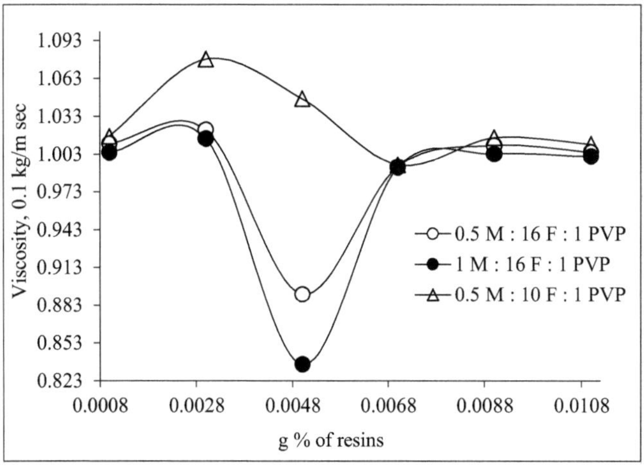

Figure 61.2 Viscosity of melamine–formaldehyde–polyvinylpyrrolidone (MFP) polymer resin with its increasing percentage composition in its (MFP) aqueous solutions. The R_1, R_2, and R_3 represent variable stoichiometric compositions of melamine, formaldehyde, and polyvinylpyrrolidone in MFP resins, respectively. Overlapping of viscosity data for the three resin stoichiometries is observed.

better resolution in data whereas when only FFs are considered in the case of viscosity then overlapping points are noticed. It infers that friccohesity improves the resolution of the data and gives better explanations of the solution-solvent interactions.

In fact friccohesity takes into account the CFs and the intermolecular forces (IMFs) or the IMFs of the solvent, which remains unengaged, and also the CFs of the solvent, which remains unengaged during its interactions with solute or the dispersate. Both Figs. 61.1 and 61.2 depict such a mechanism because in calculation for friccohesity both CFs as well as IMFs are incorporated contrary to the viscosity, where only the IMF is considered in calculation.

Since the viscous flow and drop size are affected by thermal energy, electrical energy, and infrared (IR) rays the measurements could be repeated under above-said conditions with different geometrical dimensions of the capillaries for tracking the response of the molecules. The results could be discussed and their impact could be explored for structural modifications for suitability in industrial applications as a response of UV and visible ray light, temperature, and moistures to the resultant materials.

Chapter 62

Designing a New Survismeter: Experiments

Currently survismeter science is booming in areas of drug designing, oil industries, dyeing, graphene inks, thin film and coating, capping industries, wettability for pesticides, sprayants, paints, varnishes, pigments, cosmetics, and many others. Recently, the Rivers State University of Science and Technology, Nigeria, conducted a three-day workshop on the science and operation of the survismeter. The Global-Survismeter Research Group (G-SRG) conducted workshops in Indonesia, Poland, Taiwan, and Nepal. An exponentially increasing interest in survismeter science and applications of the survismeter is proving a great laboratory asset for educating people in the form of joint academic as well as industrial meets/workshop. The G-SRG keeps organizing workshops on physicochemical interfaces for academic and industrial applications of nanoemulsions. A basic understanding and quantitative analysis of these physicochemical parameters is essential for formulating academic as well as industrial liquid mixtures. The chemical distribution and homogenization of the constituents need backgrounds of molecular forces vis-a-vis cohesive forces (CFs), intermolecular forces (IMFs), hydrophobics of constituents, van der Waals forces, Lennard-Jones potential

Survismeter: Fundamentals, Concepts, and Applications
Man Singh and Sunita Singh

ISBN 978-981-4774-70-3 (Hardcover), 978-0-429-02761-1 (eBook)
www.panstanford.com

(LJP), London dispersive forces (LDPs), and others, which are key factors for the quality of the final products. Such scientific interfacial simulation enhances clarity about the theories and principle of molecules for better application and development of new formulations, especially nanoemulsions. The survismeter tracks these parameters together. For understanding the concept of liquid mixtures, especially industrial formulations based on nanoemulsions, and developing a deep understanding of the above-mentioned conceptual sciences, a basic knowledge of these physicochemical properties (PCPs) is helpful, specifically when designing quality products of cosmetics, soaps and detergents, sprayants, ink, dyeing, thin films, and air–water bubbles (AWBs).

For educating the learners, industrial entrepreneurship, and other interested sectors, academic-cum-industrial training on academic-cum-industrial formulations is needed.

62.1 Survismeter Science and Applications

Interfacing and consolidating the molecular expressions in academic and industrial processes,

The survismeter is a green laboratory device for measuring:

- Dynamic surface tension (mN/m) or surface energy (mJ/cm^2)
- Liquid–liquid interfacial tension (LL-IFT, mN/m)
- Dynamic viscosity (indicator of solute distribution chemically)
- Kinematic viscosity and nanoemulsions
- Wettability and nanoemulsions
- Surface area and particle size of nanoemulsions
- Friccohesity and $-CH_2-$ group contribution to alkyl chains of surfactants
- Micellar studies of surfactants with salts
- Activation energy of Newtonian and non-Newtonian liquids

62.2 Installation, Operation, and Experimentation

The first step in installation is to open the packing box and vertically mount the survismeter on the saddle groove made in the base of the thermostat. Read the instructions for installation carefully and note the list of accessories attached to the packed box.

(1) Always keep the top of the box facing the sky at a 90° angle.
(2) Cautiously unpack the hardboard packing.
(3) Take out the survismeter unit.
(4) Ensure the number of stoppers is four.
(5) The box will have the following items:

- An L-shaped cone that assists in lifting the sample
- A silicon tube (a transparent and soft tube)
- An IFT (interfacial tension) capillary
- An electronic timer/racer
- A manual pendant drop number (pdn) counter
- A light reflector and a thermostat
- A leveler in general in-built on top of the thermostat
- A survismeter unit with a Sachin plate
- A piston pump (syringe)
- A thermometer to note the temperature
- A light reflector for viewing the capillary rise of the sample in the IFT capillary
- An IFT capillary fixed in the major limb (ML) of the survismeter

62.2.1 Mounting the Survismeter on the Groove of the Saddle at the Base of the Thermostat

(1) Hold the survismeter cautiously by both hands.
(2) Before sample filling in the survismeter check its proper fitting.
(3) Ensure the survismeter is in a perfectly vertical position.

62.2.2 Cleaning and Drying the Survismeter

(1) The survismeter is cleaned with Millipore water followed by acetone. Chromic acid could also be used for absolute cleaning, followed by Millipore/distilled water and acetone.
(2) The survismeter is dried by keeping it in an oven at 90°C to 110°C.
(3) The survismeter is horizontally kept on the oven's heating surface.

62.2.3 Sample Filling in the RB

(1) Fill the sample through the ML of the survismeter. The ML is connected to the reservoir bulb (RB) of the survismeter.
(2) Ensure proper leveling of the sample in the RB.
(3) Ensure two-thirds of the RB is filled with the sample.
(4) One-third of the volume should be left for aerodynamic balances.
(5) For filling the sample the survismeter unit should be tilted at a 45° angle.
(6) Surfactant containing solutions are smoothly added to avoid froth. A Teflon capillary fitted with a needle and a syringe could be used for adding surfactant solutions.

62.2.4 Prerequisites for Accurate Results

- No air should be trapped in the sample.
- The sample should not be jerked.
- A thermodynamically and kinetically stable sample should be added.
- An equilibrated sample should be subjected for measurements.
- The survismeter's position should not be changed during measurements.
- The temperature should be constant as it induces molecular motions.
- No overhead pressure should be changed during measurements.
- No disturbance or shaking should happen during measurements.

62.2.5 Hypothesis and Theme for Conducting Experiments

Several experiments could be designed and conducted with the survismeter. The systems could be made choosing the following formulations:

- Designing and formulating liquid mixtures for academics
- Molecular activities with a solvent as the medium
- Ionic activities with a solvent

- Molecular and ionic activities in biomolecules with cationic, anionic, and nonionic surfactants, ionic liquids, webdrimers
- Similar solvent activities
- Dissimilar solvents activities
- Compositional variations on the solute scale
- Constitutional variations on the solute scale

Chapter 63

Temperature-Controlled Interactions

Variations in temperature affect molecular and ionic interactions individually and also when both molecules and ions are taken together with similar and dissimilar solvents. Either the ion–dipolar interactions are largely affected, or the molecule–dipolar interactions are largely affected. Ion–dipolar interactions develop a stronger ionic field near the ion and hence seem to cause stronger ion–dipolar interactions. Therefore, molecular and ionic interactions engineering is interfaced with measurements made with the survismeter. The temperature induces atomic oscillations where the adjoining bonds also tend to oscillate to compensate the molecules as per Le Chatelier's law. Thereby, the temperature must remain constant during measurements.

63.1 Tracking the Science of Liquid Mixture Formulations

Liquid mixture formulations (LMFs) cover overall chemical, ionic, and physical sciences for attaining the desired activities of liquid mixtures, like washing, cleaning, immobilization, salting out or salting in, nanoparticle (NP) capping, coating, thin film forming,

Survismeter: Fundamentals, Concepts, and Applications
Man Singh and Sunita Singh

ISBN 978-981-4774-70-3 (Hardcover), 978-0-429-02761-1 (eBook)
www.panstanford.com

coloring, dissociation or association, dimerization-aggregation-coalescence, coagulation, encapsulations, heating, cooling, drug binding or release, DNA intercalations, graphene exfoliation, single-crystal nucleation, catalysis, adsorption, desorption, tissue repair, bacterial immobilization, dyes, cosmetics, separation, and ionizations. Therefore, design or formulation of liquid mixtures and measurement and interpretation of experimental data for developing new themes or scientific models becomes a most fascinating chapter of the survismeter. Thhe following models of physicochemical sciences are used.

- Primary chemical processes like dissolution, dispersion, precipitation, and color change
- Secondary processes like in situ redox (Mn^{3+} acts as an auto indicator)
- Indicators for color changes
- Heat indicators and friccohesity
- Iodine adsorption by starch
- Acetic acid adsorption by activated charcoal

These activities are developed to generate extra potential of mixtures for performing functions in, for example, biological, biochemical, biotechnological, and biophysical areas.

63.2 Molecular Forces

Overall the following molecular forces keep working:

- London dispersive forces (LDFs)
- Quadrupoles (symmetry) and permanent multipoles as (Willem Hendrik) Keesom interactions and forces originating from the attraction between permanent dipoles
- Friccohesity and tentropy of macromolecules
- Intermolecular multiple forces theory (IMMFT) and localized activities of macromolecules
- Dissipating and encircling the substantial effect

- Pseudo boundaries vis-à-vis molecular aggregation affected by T, P, pH, size of ions, and molecular size and their combinatorial mechanism

Similarly, the interacting activities of organic, organometallic complexes, and biopolymers molecules involve the primary, secondary, and tertiary mechanism noted as ion-dipole, induced ion-dipole, and dipole-dipole interactions. These activities affect the chemical processes and expressions through their structural charges, which are tracked by survismeter measurements noted as follows:

- Dissolution of substances in solvent or buffer solvents
- Dispersion of surfactants and dissolution of proteins
- Dyeing processes developing chemical bonds
- Ink wettability by increasing friccohesity
- Color fading: Photons weaken binding forces or intermolecular forces (IMFs)
- Color intensity and color gaining on redox
- Color change versus micellar mechanism
- Color change versus molecular degrading
- Light matter interaction

63.3 Interacting Sites for Expressing Interacting Activities

The formulation of thermodynamically and kinetically stable liquid mixtures is the most sensitive mechanism. It can be made economical and accurate by selecting the interacting chemical species or ingredients in 1:1 stoichiometric ratios. Therefore, to ensure the quality of academic experimental verification and industrial product formulations, the modulation and optimization of their interacting activities are needed. Expertise in LMFs includes knowledge of operational activities of molecules. Molecules have interacting sites that express their interacting activities with other molecules through:

- Van der Waals forces
- LDF

- Lennard-Jones potential (LJP)
- Philic-phobic initiation
- Micellar formulation constitutional inductive initiation
- Electrostatic forces for interactions
- Pi-conjugation for optimization
- Double-bond activities
- Functional group activities
- Highest occupied molecular orbital (HOMO) and lowest unoccupied molecular orbital (LUMO) mechanisms
- Hydrophobic and hydrophilic mechanism
- Twisted wave function of electrostatic pole electronegative atom
- Normal wave function of sigma bond vis-à-vis alkyl chain

The survismeter measures the overall molecular and ionic interaction engineering responsible for accurate product formulations where the friccohesity of their aqueous solutions and applications could detect the physicochemical state of the following:

- Nanoemulsion formulation of a saliva solution of biophysical interest
- Nanoemulsion formulations of urine to check jaundice
- Eye drop for the study of its fluidity

To study the aqueous solutions and formulation of nanoemulsions of chemical materials used daily in academic and industrial applications requires a basic understanding of their interaction mechanism in context of:

- Several categories of cosmetics
- Saturated and unsaturated oils and fats
- Thin film formation of nail polish
- Philic-phobic thermodynamics
- Interfacial energy for nanoemulsion formulation
- Ionic mechanism to influence micelles
- Micellar mechanism to detect the Mn^{3+} furnished by $KMnO_4$

Liquid mixtures are capable of generating efficient fluid dynamics operationalized by their friccohesity. The constitutional makeup of friccohesity and tentropy has evolved to develop the specific molecular frameworks and to read their signatures and sensitivity with respect to temperature, pressure, concentrations, pH, and environment chemical environment in approach organic concept like

- Organic solution prepared from natural products
- Organic pesticide prepared from natural products
- Organic detergents prepared from natural products
- Organic fertilizers prepared from natural products
- Organics soap and detergents prepared from natural products

The survismeter scans and responds to activities and expressions of each component of structural constitutions of molecules at the nanoscale. The expressions could align as arrays or ensembles on optimization of similar, dissimilar, or optimization state to intensify the impact in terms of dynamics. Thus measurements with the survismeter explain the molecular changes that develop from molecular activities as a drug reaction and others and so need attention for accuracy, resolution, reproducibility, calibration and standardization, and standard deviation and scattering in data.

63.4 Wettability Experiment for Making a Conducting Sheet

Experiment: Make a NaCl ionic layer in acetone (1% to 40% at an interval of 5%) + water mixtures for making a conducting surface or a soft sheet of cloth.

Procedure: Prepare the mixtures of said acetone compositions and measure their friccohesity and wettability using the interfacial tension (IFT) capillary. Spread each sample on a solid surface of a 30 cm × 30 cm sheet of cloth or filter paper. Evaporate the mixture so that the solvent can be removed and a dried layer can be obtained. The electrode of a sensor, such as ammeter, is brought in contact with the dried layer on which the ions are present spreaded. The layer conducts the current.

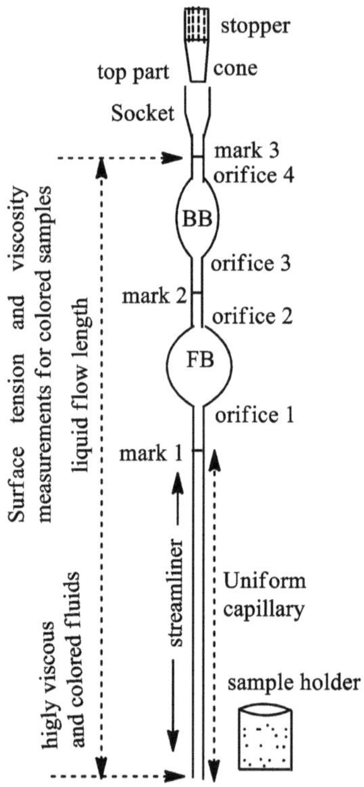

Figure 63.1 Labeled diagram of capillary used for measuring interfacial tension. The sample if filled from the open end at the bottom of the capillary. The sample is lifted up the capillary by evacuating air by using the piston pump and, in turn, reducing pressure inside the FB and BB.

Enabling Adsorption of More Gases in Nanoemulsions

Theory

Oil/water nanoemulsions with several acoustic molecules, like a series of dendrimers from trimesoyl tridimethyl malonate ester (TTDMM) to trimesoyl tridihexyl malonate ester (TTDHM), could be prepared where larger amounts of gases could be adsorbed. The dendrimers are monodispersed and also have larger void spaces where gases like CO_2, CO, NO_x, and SO_x are adsorbed in larger amounts through weak van der Waals forces.

Oxygen Adsorption by Curcumin Nanoemulsions

A curcumin nanoemulsion is prepared by dispersing it with oil-water-surfactants-ethanol-glycerol. It develops a stable nanoemulsion that can adsorb O_2 and other gases. The gas-loaded nanoemulsion could be investigated with its friccohesity values measured with the survismeter.

Friccohesity Measurements for a Colored Sample: Ink, Dye, and Indicator

Sometimes, it is difficult to read the column head of the flowing colored solution samples vis-à-vis marks made on the capillary of the survismeter. So it becomes difficult to notice the back and the upward liquid flows of the colored samples. In such a situation the colored sample, like ink or dye, is taken up to the upper marks and the whole sample is allowed to come out of the functional bulb (FB) so the pendant drop number (pdn) is measured and similarly the vft is measured (Fig. 63.1). The IFT capillary is depicted in the figure, which is self-explanatory and is leveled with needful action. The liquid lifting piston through the silicon soft tube fitted with the U-shaped cone and syringe could be used by fitting its cone in the socket of the IFT capillary.

Bibliography

1. Singh M: A simple instrument for measuring the surface tension and viscosity of liquids. *Instruments and Experimental Techniques* 2005, 48(2): 270–271.
2. Singh M: Survismeter–type I and II for surface tension, viscosity measurements of liquids for academic, and research and development studies. *Journal of Biochemical and Biophysical Methods* 2006, 67: 151–161.
3. Singh M, Patel JS, and Kale RK: Thermodynamics of philicphobic interaction shift in aqueous tweens 20–80, *International Journal of Thermodynamics* 2011, 14: 135–146.
4. Singh M and Irshad Md: Study of aqueous biomolecules with magnetic flux using density, viscosity, and surface tension probe at 304.15 K. *Journal of Dispersion Science and Technology* 2008, 29: 895–898.
5. Singh M and Kumar V: Solvodynamics of benzene and water phases by DTAB, MTOAC, TMSOI and orcinol studied with interfacial tension, surface tension and viscosity measured with survismeter. *International Journal of Thermodynamics* 2008, 11(4): 181–186.
6. Pandya SR and Singh M: Preparation and characterization of magnetic nanoparticles and their impact on anticancer drug binding and release processes moderated through a 1st tier dendrimer. *RSC Advances* 2016, 6(44): 37391–37402.
7. Singh M, Kumar V, Kale RK, and Jain CL: Molecular activation energies (delta mu(2)*) of L-lysine, L-tyrosine, L-proline, DL-alanine, glycerol, orcinol, iodine, DTAB, and TMSOI for blending with melamine-formaldehyde-polyvinylpyrrolidone polymer resin illustrated with SEM. *Journal of Applied Polymer Science* 2010, 118(2): 960–968.
8. Hussan R: Nanoemulsion as a novel transdermal drug delivery system. *International Journal of Pharmaceutical Sciences and Research* 2011, 2(8): 1938–1946.
9. Sharma N, Bansal M, Visht S, et al.: Nanoemulsion: A new concept of delivery system. *Chronicles of Young Scientists* 2010, 1(2): 2–6.

10. Bhatt P and Madhav S: A detailed review on nanoemulsion drug delivery system. *International Journal of Pharmaceutical Sciences and Research* 2011, 2(10): 2482–2489.
11. Quintero L, Mckellar JA, and Clark DE: Nanoemulsion. Patent no. US8822385B2, 2014.
12. Akhter A, Farhan J, Jain K, et al.: Investigation of nanoemulsion system for transdermal delivery of domperidone: Ex-vivo and in vivo studies. *Current Nanoscience* 2008, 382.
13. Devarajan V and Ravichandran V: Nanoemulsions—As modified drug delivery tool. *International Journal of Comprehensive Pharmacy* 2011, 4 (01): 1–6.
14. Pershing LK, Lambert LD, and Knutson K: Mechanism of ethanolenhaced estradiol permeation across human skin in vivo. *Pharmaceutical Research* 1990, 7: 170–175.
15. Shah P and Bhalodia D: Nanoemulsion—A Pharmaceutical Review. *Systematic Reviews in Pharmacy* 2010, 1(1): 24–32.
16. Liu P, Kurihara-Bergstrom T, and Good WR: Cotransport of estradiol and ethanol through human skin in vitro: understanding the permeant/enhancer flux relationship. *Pharmaceutical Research* 1991, 8: 938–944.
17. Kim YH, Ghanem AH, Mahmoud H, et al.: Short chain alkanols as transport enhancers for lipophilic and polar/ionic permeants in hairless mouse skin: mechanism(s) of action. International Journal of Pharmaceutics 1992, 80: 17–31.
18. Pershing LK, Parry GE, and Lambert LD: Disparity of in vitro and in vivo oleic acid-enhanced b-estradiol percutaneous absorption across human skin. *Pharmaceutical Research* 1993, 10: 1745–1750.
19. Tanojo H, Junginger HE, and Boddé HE: In-vivo human skin permeability enhancement by oleic acid: transepidermal water loss and Fouriertrans-form infrared spectroscopy studies. *Journal of Control Release* 1997, 47: 31–39.
20. Gupta P and Pandit J: Pharmaceutical nanotechnology novel nanoemulsion–high energy emulsification preparation, evaluation and application. *The Pharma Research* 2010, 3: 117–138.
21. Simonnet et al.: Nanoemulsion based on ethoxylated fatty ethers or on ethoxylated fatty esters and its uses in the cosmetics, dermatological and/or ophthalmological fields. Patent US6375960B1, 2002.
22. Hadgraft J: Skin, the final frontier. *International Journal of Pharmaceutics* 2001, 224: 1–18.

23. Fang JY, Yu SY, Wu PC, et al.: In-vitro skin permeation of estradiol from various proniosome formulations. *International Journal of Pharmaceutics* 2001, 215: 91–99.
24. Kumar M, Mishra A, Pathak K, et al.: Intranasal nanoemulsion based brain targeting drug delivery system of risperidone. *International Journal of Pharmaceutics* 2008, 358: 285–291.
25. Ruktanonchai U, Sakulku U, Nuchuchua O, et al.: Characterization and mosquito repellent activity of citronella oil nanoemulsion. *International Journal of Pharmaceutics* 2009, 112: 256–282.
26. Tajima K, Imai Y, Ujiie N, et al.: Three phase structure of hexadecane nanoemulsion formed with phospholipid-surfactant mixtures and its novel Phase transition temperature TE colloids and surfaces. *Pharmaceutical Research* 2006, 276: 134–142.
27. Basphinar Y, Keck MC, and Borchert HH: Development of a positively charged prednicarbate nanoemulsion. *International Journal of Pharmaceutics* 2010, 383: 201–208.
28. Junyaprasert BV, Muller HR, Souto BE, et al.: Q10 loaded NLC versus nanoemulsions; stability, rheology and in vitro skin permeation. *International Journal of Pharmaceutics* 2009, 377: 207–214.
29. Constantinides PP, Chaubal VM, and Shorr R: Advances in Lipid nanodispersions for parental drug delivery and targeting. *Advanced Drug Delivery Reviews* 2008; 60: 757–767.
30. Vyas TK, Amiji MM, and Shahiwala A: Improved oral bioavailability and brain transport of saquinavir upon administration in novel nanoemulsion formulations. *International Journal of Pharmaceutics* 2008, 347: 93–101.
31. Leong TSH, Kentish SE, Wooster TJ, et al.: Minimizing oil droplet size using ultrasonic emulsification. *Ultrasonics Sonochemistry* 2009, 16: 721–727.
32. Aubrun OS, Simonnet JT, and Alloret FL: Nanoemulsions: a new for skincare products. *Journal of Colloid and Interface Science* 2004, 108–109: 145–149.
33. Nicolosi JR, Kuo F, Kotyla T, et al.: Nanoemulsion of an anti-oxidant synergy formulation containing gamma tocopherol have enhanced bioavailability and anti-inflammatory properties. *International Journal of Pharmaceutics* 2008, 363: 206–213.

Index